ASCE/EWRI 44-05

American Society of Civil Engineers

Standard Practice for the Design and Operation of Supercooled Fog Dispersal Projects

This document uses both the International System of Units (SI) and customary units.

Published by the American Society of Civil Engineers

Library of Congress Cataloging-in-Publication Data

American Society of Civil Engineers.
Standard practice for the design and operation of supercooled fog dispersal projects / Atmospheric Water Management Standards Committee, Environmental & Water Resources Institute ; subcommittee membership, Thomas P. DeFelice (chair/editor) ... [et al.].
p. cm.—(ASCE standard)
"ASCE/EWRI 44-05."
Includes bibliographical references.
ISBN 0-7844-0795-9
1. Fog—Control. 2. Precipitation (Meteorology)—Modification. I. DeFelice, Thomas P., 1959- II. Environmental and Water Resources Institute (U.S.). Atmospheric Water Management Standards Committee. III. Title IV. Series: American Society of Civil Engineers. ASCE standard.

QC929.F7A536 2005
551.68'75—dc22

2005000564

The material presented in this publication has been prepared in accordance with generally recognized engineering principles and practices and is for general information only. This information should not be used without first securing competent advice with respect to its suitability for any general or specific application.

The contents of this publication are not intended to be and should not be construed to be a standard of the American Society of Civil Engineers (ASCE) and are not intended for use as a reference in the purchase of specifications, contracts, regulations, statutes, or any other legal document.

No reference made in this publication to any specific method, product, process, or service constitutes or implies an endorsement, recommendation, or warranty thereof by ASCE.

ASCE makes no representation or warranty of any kind, whether express or implied, concerning the accuracy, completeness, suitability, or utility of any information, apparatus, product, or process discussed in this publication and assumes no liability therefore.

Anyone utilizing this information assumes all liability arising from such use, including but not limited to infringement of any patent or patents.

ASCE and American Society of Civil Engineers—Registered in U.S. Patent and Trademark Office.

Library of Congress Catalog Card No: 2005000564
ISBN 0-7844-0795-9
Manufactured in the United States of America.

STANDARDS

In April 1980, the Board of Direction approved ASCE Rules for Standards Committees to govern the writing and maintenance of standards developed by the Society. All such standards are developed by a consensus standards process managed by the Codes and Standards Activities Committee. The consensus process includes balloting by the Balanced Standards Committee, which is composed of Society members and nonmembers, balloting by the membership of ASCE as a whole, and balloting by the public. All standards are updated or reaffirmed by the same process at intervals not exceeding 5 years.

The following Standards have been issued:

ANSI/ASCE 1-82 N-725 Guideline for Design and Analysis of Nuclear Safety Related Earth Structures
ANSI/ASCE 2-91 Measurement of Oxygen Transfer in Clean Water
ANSI/ASCE 3-91 Standard for the Structural Design of Composite Slabs and ANSI/ASCE 9-91 Standard Practice for the Construction and Inspection of Composite Slabs
ASCE 4-98 Seismic Analysis of Safety-Related Nuclear Structures
Building Code Requirements for Masonry Structures (ACI 530-02/ASCE 5-02/TMS 402-02) and Specifications for Masonry Structures (ACI 530.1-02/ASCE 6-02/TMS 602-02)
SEI/ASCE 7-02 Minimum Design Loads for Buildings and Other Structures
ANSI/ASCE 8-90 Standard Specification for the Design of Cold-Formed Stainless Steel Structural Members
ANSI/ASCE 9-91 listed with ASCE 3-91
ASCE 10-97 Design of Latticed Steel Transmission Structures
SEI/ASCE 11-99 Guideline for Structural Condition Assessment of Existing Buildings
ANSI/ASCE 12-91 Guideline for the Design of Urban Subsurface Drainage
ASCE 13-93 Standard Guidelines for Installation of Urban Subsurface Drainage
ASCE 14-93 Standard Guidelines for Operation and Maintenance of Urban Subsurface Drainage
ASCE 15-98 Standard Practice for Direct Design of Buried Precast Concrete Pipe Using Standard Installations (SIDD)
ASCE 16-95 Standard for Load and Resistance Factor Design (LRFD) of Engineered Wood Construction
ASCE 17-96 Air-Supported Structures
ASCE 18-96 Standard Guidelines for In-Process Oxygen Transfer Testing
ASCE 19-96 Structural Applications of Steel Cables for Buildings
ASCE 20-96 Standard Guidelines for the Design and Installation of Pile Foundations
ASCE 21-96 Automated People Mover Standards—Part 1
ASCE 21-98 Automated People Mover Standards—Part 2
ASCE 21-00 Automated People Mover Standards—Part 3
SEI/ASCE 23-97 Specification for Structural Steel Beams with Web Openings
SEI/ASCE 24-98 Flood Resistant Design and Construction
ASCE 25-97 Earthquake-Actuated Automatic Gas Shut-Off Devices
ASCE 26-97 Standard Practice for Design of Buried Precast Concrete Box Sections
ASCE 27-00 Standard Practice for Direct Design of Precast Concrete Pipe for Jacking in Trenchless Construction
ASCE 28-00 Standard Practice for Direct Design of Precast Concrete Box Sections for Jacking in Trenchless Construction
SEI/ASCE/SFPE 29-99 Standard Calculation Methods for Structural Fire Protection
SEI/ASCE 30-00 Guideline for Condition Assessment of the Building Envelope
SEI/ASCE 31-03 Seismic Evaluation of Existing Buildings
SEI/ASCE 32-01 Design and Construction of Frost-Protected Shallow Foundations
EWRI/ASCE 33-01 Comprehensive Transboundary International Water Quality Management Agreement
EWRI/ASCE 34-01 Standard Guidelines for Artificial Recharge of Ground Water
EWRI/ASCE 35-01 Guidelines for Quality Assurance of Installed Fine-Pore Aeration Equipment
CI/ASCE 36-01 Standard Construction Guidelines for Microtunneling
SEI/ASCE 37-02 Design Loads on Structures During Construction
CI/ASCE 38-02 Standard Guideline for the Collection and Depiction of Existing Subsurface Utility Data
EWRI/ASCE 39-03 Standard Practice for the Design and Operation of Hail Suppression Projects
ASCE/EWRI 40-03 Regulated Riparian Model Water Code
ASCE/EWRI 42-04 Standard Practice for the Design and Operation of Precipitation Enhancement Projects
ASCE/SEI 43-05 Seismic Design Criteria for Structures, Systems, and Components in Nuclear Facilities
ASCE/EWRI 44-05 Standard Practice for the Design and Operation of Supercooled Fog Dispersal Projects

FOREWORD

The ASCE Board of Direction approved a revision to the ASCE Rules for Standards Committees in April 1995 to govern the writing and maintenance of standards developed for the society. A consensus process managed by the ASCE Codes and Standards Activities Committee (CSAC) develops such standards. The consensus process includes balloting by a balanced ASCE Environmental and Water Resources Institute (EWRI), Atmospheric Water Management (AWM) Standards Committee (SC), made up of Society members and nonmembers, balloting by the membership of ASCE as a whole and the public. All standards are updated or reaffirmed by the same process at intervals not to exceed five years.

The provisions of this document have been written in permissive language and, as such, offer a series of options or instructions to the user, but do not prescribe a specific course of action. Significant judgment has been left to the user of this document.

This document describes the process through which supercooled fog dispersal operations should be designed and conducted. This document provides information on the planning, conduct, and evaluation of such efforts. Its intended audience will not have significant expertise in weather modification science, especially as pertains to supercooled fog dispersal. As a result, a technical section on fog characteristics has been added. This technical section empowers these people with the knowledge most critical to conducting a successful supercooled fog dispersal project.

This standard has been prepared in accordance with the CSAC style manual, 4 September 1998 revision, with recognized engineering principles and should not be used without the user's competent knowledge of the underlying principles for a given application. The International System of Units (SI units) is used throughout, with English equivalents also provided. Exceptions are the use of the Celsius (°C) temperature scale and, where appropriate, centimeters (cm) in lieu of m. Section 8 provides all conversion factors used herein. Italics denote special emphasis or a glossary entry.

The publication of this standard by ASCE is not intended as warrant that the information contained therein is suitable for any general or specific use, and the Society takes no position respecting the validity of patent rights. The user is advised that the determination of patent rights or risk of infringement is entirely his or her own responsibility.

Many have contributed materially to this document by their comments, review, illustrations, and photographs. The primary authors of this document were the EWRI AWM SC Fog Dispersal Subcommittee Members: Thomas P. DeFelice (Chair/Editor), Maurice Roos, Conrad G. Keyes Jr., Richard H. Stone III, and Thomas J. Henderson. Others who have contributed materially include Bruce A. Boe, George W. Bomar, Donald A. Griffith, James Heimbach Jr., Gary Riley, James Renick, and Harold D. Orville.

Thomas P. DeFelice, Ph.D.
Raytheon ITSS
Upper Marlboro, MD 20774
Thomas_P_DeFelice@Raytheon.com

ACKNOWLEDGMENTS

The American Society of Civil Engineers (ASCE) recognizes the work of the Atmospheric Water Management Standards Committee of the Environmental and Water Resources Institute (EWRI), in particular Subcommittee Members Thomas P. DeFelice (Chair/Editor), Maurice Roos, Conrad G. Keyes, Jr., Richard Stone, and Thomas J. Henderson. We also acknowledge the many who contributed their comments, reviews, illustrations, and photographs.

CONTENTS

Design and Operation of Supercooled Fog Dispersal Projects

1.0 INTRODUCTION

Fogs can pose a significant threat to public safety and quality of life in the air, on land, and at sea. For example, the luxury liner *Andrea Doria* collided in fog off New York with the *Stockholm* on its 1956 maiden voyage and sank. Fifty-one people died and millions of dollars in property was lost (Silverman and Weinstein, 1974). Extended periods of fog can have large economic impacts on the aviation, tourism, transportation, and mining industries. For example, the loss of revenue at one airport, by one fog, because of aircraft diversions, delays, and cancellations was estimated by the Department of Transportation as approximately $100,000 in the early 1970s (Silverman and Weinstein, 1974). These extended fog periods can have negative impacts on agriculture and the mental health of the general public. In contrast, there are some situations in which fog is beneficial, such as where fog water is collected for drinking water in arid regions (e.g., Schemenauer, 1998) and where fog supplies some of the necessary moisture to vegetation. For example, fog supplies a necessary part of the moisture needed by the northern California redwood trees during the summer dry season (e.g., Schemenauer, 1998). Another example is the notorious winter fog in the San Joaquin Valley of California, which provides an important portion of the winter dormancy requirements of many deciduous orchard crops in the region. The San Joaquin Valley fogs are also known as "tule" fogs and are not normally supercooled as temperatures are often just above freezing.

The importance of transportation alone was sufficient justification for attempts to modify or disperse fogs. Silverman and Weinstein (1974) note that fog was the subject of the first scientifically designed weather modification effort of any kind. This may partially explain why supercooled fog dispersal is perhaps the only weather modification technology that does not require long experimentation and careful measurement to detect results, since results are both visible and nearly instantaneous. The most frequently cited goal of any supercooled fog dispersal project is to increase visibility. An increase in the local temperature can be a by-product of the clearing activities. Fog dispersal operations reduce the threat to public safety by increasing the visibility over highways and airport runways. Dispersing fog to increase visibility, especially at airports, has tremendous economic value, particularly at the local level, as transportation returns to normal levels. Additional sunshine resulting from fog dispersal operations can often increase the quality of life for specific populations. Fog clearing in open pit mines can allow the safe resumption of mining operations that were previously suspended as a result of decreased visibility.

Ice fogs are a special case and are naturally slower to dissipate than supercooled fogs, since they are composed mostly of tiny ice crystals, and they generally form at air temperatures below about −30°C (e.g., Huffman and Ohtake, 1971). Ice fog dispersal is fundamentally different from the dispersal of supercooled fogs and may more appropriately be labeled ice fog suppression. Ice crystals predominate and form by *heterogeneous nucleation* and in some instances by *homogeneous nucleation*. The existence of ice fogs is primarily due to unnatural sources of water vapor, which may include automobile and aircraft exhaust, exhaust from utility plants, and open water (e.g., cooling ponds). Benson (1969) indicated that decreasing the ambient temperature of these moisture sources did improve visibility. Most attempts to disperse ice fogs have included electric fields, dehydrators of various types (gas, furnace, automobile, etc.), air movement by helicopters, polyethylene rafts, plastic films (e.g., polyethylene), injection wells, cooling towers, and chemical films (e.g., hexadeconal, ethylene glycol monobutyl ether). Presently, the standard technique used to suppress ice fog, caused by exposed water sources, employs a thin ethylene glycol monobutyl ether film. The ethylene glycol monobutyl ether film (1) is harmless to marine life (i.e., biodegradable), (2) is less effective in suppressing ice fog than hexadeconal, and (3) lasts much longer than other films. McFadden (1976) and McFadden and Collins (1978) provide details of these techniques. Ice fog suppression techniques will not be discussed in this document. The focus of this standard practice document is on the dispersal of supercooled fogs.

1.1 HISTORICAL REVIEW OF SUPERCOOLED FOG DISPERSAL OPERATIONS

Supercooled fog is a stable *colloid*, but is otherwise in a thermodynamically metastable state (e.g., Silverman and Weinstein, 1974). Thus, supercooled fog can be dissipated by growth and sedimentation of ice crystals. Seeding supercooled fog with ice-forming particles or *nuclei* may yield visibility improvements

FIGURE 1-1. Long rectangular hole cut through a stratus cloud using crushed dry ice at the rate of about 1 kg/km.

of at least a mile within 15 minutes following seeding (Figure 1-1). The results are so repeatable that randomized statistical verification is generally considered unnecessary. As a result, supercooled fog dispersal has been operational since 1950 in the United States and since 1952 in Russia. The most frequent locations of these operations are at airports. Suitable seeding techniques are primarily dependent on wind, temperature, and supercooled liquid water amount. Some involve the introduction of artificial ice *nuclei* into the air from either ground-based or airborne delivery systems. Other techniques employ liquid carbon dioxide, nitrogen, or propane to disperse fogs. Liquid nitrogen is used in ground-based systems at a few airports in some countries (WMO, 1992).

Fog-clearing operations are required under U.S. law to be reported to the National Oceanic and Atmospheric Administration (NOAA). However, it would appear that some supercooled fog dispersal projects may not be reported to NOAA. For example, there were nine fog dispersal activities reported during 1973 (Charak, 1981), whereas there were three each year between 1991 and 1993 according to Blackmore (1992, 1994). No fog activities were reported to NOAA during 1998. These numbers do not necessarily reflect a sharp reduction of fog dispersal activities. Sponsors should periodically publish the results of these activities, since knowing about them could improve our understanding about fogs and their impact on society and the environment.

The remainder of this document includes the capability statements for fog dispersal, the abridged version of the physics of supercooled fog formation and dispersal, as well as how one might plan, organize, conduct, and evaluate a supercooled fog dispersal operation.

1.2 THE STATUS OF SUPERCOOLED FOG DISPERSAL TECHNOLOGY

Since 1964, when the National Science Board appointed a Special Commission of Weather Modification, chaired by Dr. A.R. Chamberlain, found that *"supercooled fog on the ground can be dissipated. . . ."* (e.g., NRC, 2003), policy and/or capability statements on the subject of supercooled fog dispersal have been created. The following summarize the present perspectives of the scientific establishment on the subject of supercooled fog dispersal.

1.2.1 American Society of Civil Engineers

The American Society of Civil Engineers' Manual of Professional Practice No. 81 (Kahan et al., 1995) offers guidelines that cover general cloud seeding projects and techniques.

The American Society of Civil Engineers (ASCE) supports and encourages the protection and prudent development of the Nation's atmospheric water (also known as "weather modification" or "cloud seeding") for beneficial uses. Sustained support for atmospheric water data collection, research and operational programs, and the careful evaluations of such efforts including the assessment of extra-area and long-term environmental effects, is essential for prudent [such] development. ASCE recommends that the results and findings of all atmospheric water-management programs and projects be freely disseminated to the professional community, appropriate water managers, and to the public (ASCE 2003).

1.2.2 Weather Modification Association

The most recent supercooled fog dispersal capability statement adopted by the Weather Modification Association (WMA, 1986) reads:

The dispersal of shallow, cold (below freezing) fog or stratus cloud decks is an established operational technology. Dispersing ice phase seeding agents, such as dry ice or silver iodide, in these situations is effective in improving visibility. Clearings established in cloud decks, embedded in strong wind fields, fill in quickly unless seeding is done nearly continuously.

1.2.3 American Meteorological Society

The most recent supercooled fog dispersal capability statement adopted by the American Meteorological Society (AMS, 1998) reads:

Operations that dissipate cold fog and cold stratus clouds by seeding with ice-forming agents (e.g., dry ice, liquid nitrogen, compressed air, silver iodide, etc.) have become routine at some airports.

1.2.4 World Meteorological Organization

The current WMO capability statement on supercooled fog dispersal (WMO, 1992) reads:

Certain techniques have been shown to be effective in clearing warm and cold fogs. Cold fog can be dissipated by growth and sedimentation of ice crystals. This may be induced with high reliability by seeding the fog with artificial ice nuclei (IN) from ground-based or airborne systems. This technique is in operational use at several airports where there is a relatively high incidence of cold fog. Suitable techniques are dependent upon wind, temperature, and other factors. Dry ice has commonly been used in airborne systems. Other systems employ rapid expansion of compressed gas (propane) to cool the air enough to form ice crystals. At a few airports in some countries, liquid nitrogen is being used in ground-based systems. Because the effects of this type of seeding are easily measured and results highly predictable, randomized statistical verification generally has been considered unnecessary.

2.0 FOG CHARACTERISTICS

Whoever undertakes a fog dispersal project must first understand what kind of fog they are dealing with and what characteristics it possesses. For example, where and when was it formed, and what caused it to form (radiative cooling or did the fog move into the area from somewhere else)? Fogs produced at very low relative humidities (as low as 87%) typically have numerous fog droplets per unit volume (or high fog droplet number concentrations). Furthermore, fogs with high droplet concentrations not only produce low visibilities (e.g., Malone, 1951), but also are very slow to dissipate.

Fog is herein defined as a cloud whose base is at ground level. Fog has had many definitions since the mid 1900s. The international definition of fog stated that it was a cloud consisting of minute water droplets and/or ice crystals that envelops the observer and restricts the horizontal visibility to 1,000 m or less. The international definition of mist was a fog with visibility greater than 1,000 m. Mist has been used in the United States to describe drizzle, although mist droplets are actually smaller than drizzle droplets, which are smaller than rain droplets. Fog has also been defined as a cloud whose base was at or below the elevation of the observer. This has limited application, which would not apply in the instance where the observer is on a mountaintop, engulfed in cloud. The mountaintop observers would say they were in fog, while the observers at the base of the mountain would call that cloud something else, such as stratocumulus, stratus, etc. Furthermore, older references also mention smoke and haze when discussing fog. Consequently, most earlier fog studies began with an author's classification of fog, which was usually presented in terms of causation (i.e., advection-warm air over cold land, cold air over a moist surface; radiation; advection-radiation; frontal). Malone (1951), Byers (1959), and Shemenauer (1998) provide additional details.

There are three basic kinds of fogs: warm fogs, where droplets exist in air temperatures warmer than 0°C; supercooled fogs, where droplets exist in air temperatures colder than 0°C; and ice fogs, where air temperatures are typically colder than −30°C (Huffman and Ohtake, 1971). Ice fogs are analogous to dense cirrostratus clouds (Wendler, 1969) and have ice crystal sizes between 2 and 50 μm at number densities between 30 and 700 cm^{-3} (Ohtake and Huffman, 1969). Approximately 5% of all fogs in the United States are accompanied by subfreezing air temperatures (Changnon, 1975).

Fogs may be further categorized according to the physical processes responsible for their formation. A fog produced by radiative cooling of the earth's surface is called a radiation or ground fog. For example, fogs in the Central Valley of California are generally *radiation fogs*. Advection fogs are produced from the movement of warm, moist air (as in water vapor–laden air) over a surface cold enough to induce condensation above it. Fogs along the U.S. West Coast are generally advection fogs. Advection fog that subsequently moves over relatively colder ground, which continues to cool by radiation, is termed advection-*radiation fog*. Fog that forms in moist air that flows up along a hillside is called upslope fog. Fog that forms from the mixture of two different *air masses*, whether these different *air masses* are saturated or not is typically termed evaporation-mixing fog. Steam fog is an example of evaporation-mixing fog and may be seen over the thermal ponds at Yellowstone National Park, especially during the winter months (e.g., Dennis, 1980). Another example of evaporation-mixing fog could be one's breath.

The formation of fog was not routinely forecast before the middle of the twentieth century (Malone, 1951; Byers, 1959), but there has since been some improvement in this capability (e.g., Schemenauer, 1998). Most of the improvements incorporate the use of satellite imagery, radar, and lidar data with the more traditional parameters (i.e., air temperature, dew point, and wind field data) used to forecast fogs. The weather data will also be useful in identifying the type of fog that may have to be seeded. Some numerical weather prediction models developed by the National Center for Environmental Prediction (NCEP), such as the ETA and RUC (20 km), have more detailed topography and could aid fog forecasting.

2.1 FOG DROPLET CHARACTERISTICS

The formation of fog requires either (1) air cooled to its dew or ice point or (2) water vapor added to a given volume of air until that air volume reaches saturation. There are a number of ways by which either (1) or (2) can be achieved in nature, such as radiative cooling, advection of air over a water body, advection of air over a moist or dry surface of a different temperature, and evaporation-mixing. Further details are available (see, e.g., Malone, 1951; Byers, 1959; Jiusto and Lala in ASRC, 1973; Welch and Wielicki, 1986; Tonna, 1989; Schemenauer and Caselles, 1991; Houze, 1993; Wetzel et al., 1996; Heintzenberg et al., 1998; Schemenauer, 1998).

Once the vapor begins to condense onto the larger and/or more soluble condensation *nuclei,* fog droplets begin to appear. The fog droplets grow until all the available water vapor for growth is depleted. The growth (or evaporation) of the fog droplet (or ice crystal depending on the air temperature) is simply gauged by the change in droplet mass as a function of time. Small droplets (less than approximately 100 μm in diameter, d) are essentially spherical. The mass growth of such droplets by water vapor diffusion may be expressed as a function of change in droplet diameter (or radius). Other droplet-related fog parameters include the number density, also commonly referred to as the droplet concentration, n_d. That is, the number of droplets existing within a unit volume (usually a cubic centimeter, or 1 cm^3, or 10^{-6} m^3), and the fog liquid water content, w_l, conventionally expressed as the summation of droplet masses, in grams (g) within one cubic meter (m^3), e.g., 0.5 g m^{-3}. The droplet number concentration and droplet diameters contribute directly to visibility, the parameter most often used to evaluate the effectiveness of fog dispersal projects.

Fogs have characteristic droplet (or ice crystal) number densities and droplet (or ice crystal) sizes depending on the *air mass*, continental or maritime, in which they form. Fogs that form in *continental air masses* have a tendency to consist of high number densities of droplets, whereas fogs that form in *maritime air masses* have low number densities. Fog droplet diameters can range between ~1 μm and ~200 μm and their number concentrations (or number densities) can range from several thousand to about one per cubic centimeter. (Note that the average diameter of human hair is somewhere around 65 μm, and the size of one cubic centimeter (cm^3) is roughly that of a standard sugar cube used for coffee.) The droplet (ice crystal) density, drop (or ice crystal) size, and number of these that exist within a given volume of a fog are used to determine the liquid (or ice) water content of the fog. The water content of the fog is an operational parameter that should be considered when dispersing a fog. Supercooled fogs have liquid water contents that typically range between 0.01 and 0.5 g m^{-3}, but may occasionally be greater. Pruppacher and Klett (1980), Rogers and Yau (1991), and Young (1993) provide additional details.

Fog dispersal projects use seeding agents to initiate the formation of ice, a process called ice nucleation. If freezing results from the introduction of nonice substances, the process is more specifically termed heterogeneous nucleation, and the substances used are known as ice-forming *nuclei*. The temperature threshold at which an ice-forming nucleus triggers nucleation is referred to as the nucleus' activation temperature. Freezing of supercooled water is also possible in the absence of ice-forming nuclei if the fog is sufficiently cooled. Such a process is called homogeneous ice nucleation. The temperature threshold for the homogeneous ice nucleation of pure water is −40°C (e.g., Rogers and Yau, 1991). There are many sources of natural *ice n*uclei (IN), with clay soils perhaps being the most common. These natural *ice n*uclei (IN) generally have activation temperatures on the order of −12°C to −21°C (Table 2-1), although natural ice nucleation rarely occurs at temperatures warmer than this range.

The seeding agents used to disperse fogs initiate ice formation through one of four heterogeneous modes: deposition, immersion-freezing, condensation-freezing, and contact. Deposition occurs when ice crystals develop as water vapor collects directly on ice (or initially on ice-forming *nuclei*). This ice growth occurs at the expense of the surrounding supercooled water vapor. An immersion nucleus functions after immersion within a cloud droplet, followed by freezing when the droplet cools to the nucleus' activation temperature. Condensation-freezing nucleation occurs when the nucleus also acts as a cloud condensation nucleus, thereby inducing droplet formation around it upon

TABLE 2-1. Ice-Forming Nuclei and their Ice Crystal Nucleation Threshold Temperatures

Atmospheric Ice Nucleant	Threshold Temperature (°C)	Complete Activity (°C)
Pure Ice	0	0
Dry Ice (carbon dioxide)	−1	
Cholesterol	−2	NA
Bacterium (such as *Pseudomonas syringae;* bacteria in leaf mold)	−2.6	NA
1, 5-dihydroxynaphthalene	−3	NA
Silver Iodide, *AgI*	−4	−10
Metaldehyde	−5	NA
Lead Iodide, PbI_2	−6	−12
Copper Sulfide, CuS	−7	NA
Copper Oxide, CuO	−7	NA
Vaterite	−7	NA
Mercury Di-Iodide, HgI_2	−8	NA
LOAM (ND)	−8	−25
Kaolinite (silicate)	−9	NA
Cadmium Di-Iodide, CdI_2	−12	NA
Di-Iodine, I_2	−12	NA
Clay (Guilderland, NY)	−12	−25
Volcanic Ash (Rogers and Yau, 1991)	−13	NA
Halloysite	−13	NA
Vermiculite	−15	NA
Soil (Montana and WY)	−15	−27
LOAM (Oakley, KS)	−15	−32
Cinnabar	−16	NA
LOAM (Coeur D'Alene, ID)	−17	−27
Ash (Crater Lake, OR)	−17	−29
Kyranite (Al_2SiO_5)	−19	−30
Dust (Phoenix, AZ)	−19	−30
Marl (Ravena, NY)	−19	−29
Bentonite (MT)	−21	−29
Kaolin (GA)	−22	−32
Ash (Paricutin, Mexico)	−23	−30
Soil (NV)	−24	−33
DIATOMS	−30	−39
SPORES	−36	−38

Adapted from Schaefer, 1951; Rogers and Yau, 1991.
NA = Not available.

being placed in an environment near saturation. Ice formation follows as soon as the droplet cools to the nucleus' activation temperature. Contact nucleation results when the nucleus collides with a droplet already supercooled to the nucleus' activation temperature. Homogeneous nucleation may be induced by seeding with dry ice, which has an equilibrium surface temperature of −78.5°C (Mason, 1981; Vonnegut, 1981), or by venting of certain liquefied gases (e.g., propane). In the case of supercooled fogs, the two most likely modes of ice formation are contact and deposition. Additional details of ice nucleation may be found in Dennis (1980), Pruppacher and Klett (1980), DeMott et al. (1995), Götz et al. (1991), Rogers and Yau (1991), and Young (1993).

Once formed, the ice crystals may grow by continued deposition; by contact with supercooled cloud droplets that immediately freeze to the ice crystal (a process called accretion growth); and/or by aggregation, wherein ice crystals become tangled with each other, building snowflakes. The ice particles created by seeding may therefore grow sufficiently to precipitate, clearing the fog, or at least increasing the visibility within the fog, unless there is strong *advection* of water vapor–laden air to reestablish the supercooled fog.

2.2 FOG CHARACTERISTICS AS APPLIED TO FOG DISPERSAL OPERATIONS

A fog dispersal project should not be initiated without knowing the physical characteristics of fogs that affect the target area and how they form. How a fog forms provides insight into the physical characteristics of the fog, which, in turn, provides helpful insight for selecting the dispersal strategy. Physical characteristics include air temperature, relative humidity, pressure and wind field in the foggy region, predominant water phase, droplet (or ice crystal) size and number density, and possibly the chemical composition of the fog. The chemical composition could be used to help evaluate potential environmental impacts of the fog and its dispersal. Fogs produced in air containing high cloud condensation *nuclei* (CCN) concentrations are dispersed differently than fogs produced in air containing low CCN concentrations. Fogs with large numbers of droplets typically have relatively low visibilities and are also very slow to dissipate naturally (e.g., Malone, 1951; Tomasi and Tampieri, 1976a, b). These fogs make good candidates for successful fog dispersal operations. Fogs form under many atmospheric conditions, so dispersal techniques must be selected accordingly. Certain fog dispersal techniques have been shown to be effective in clearing supercooled fogs.

The goal of most fog dispersal operations is to increase visibility. Visibility, V, within a fog is inversely proportional to the geometric cross sections of the fog hydrometeors. There is usually only a small range of fog hydrometeor sizes. The numbers of droplets of any given diameter vary, depending on the available water vapor, aerosols, and formation mechanism. According to Hess (1974), visibility will be better in a supercooled fog composed of small droplets compared to one composed of large supercooled droplets, with fog droplet number concentration held constant. Furthermore, for any supercooled fog with a given uniform droplet size, visibility will be improved when these droplets exist in low concentrations. This dependence on droplet size and concentration suggests the visibility in supercooled fog is also inversely related to the supercooled water content of a fog with a mean value of droplet diameter, d. The water content, w_l, is a product of the number density, n_d, the diameter cubed of the *hydrometeor*s, and the density of water. Visibility during nighttime depends on the same parameters as daytime visibility. It is suggested that one use a commercial nighttime visibility measuring device or markers (such as a beacon on a tower) that are known distances away from one's location in conjunction with local weather data to determine nighttime visibility. Gazzi et al. (1975a, b) provide some insights into daytime and nighttime visual range determination at airports.

One can come up with scenarios wherein a cloud volume could have equivalent liquid water contents and very different droplet sizes, thus resulting in different visibilities (Table 2-2).

The real importance of Table 2-2 is realized when trying to devise a strategy for dispersing a fog. The appropriate application of seeding material can change the physical characteristics of a fog with $w_l = 0.02$ g m^{-3}, $n_d = 500$ cm^{-3}, and d = 9.2 μm into a fog with $w_l = 0.02$ g m^{-3}, $n_d = 50$ cm^{-3}, and d = 19.7 μm leading to a corresponding increase in visibility. The method that will most effectively achieve and maintain this increase in visibility depends on a number of factors, as indicated above, especially the kind of fog being seeded.

The target visibility or range of values desired for various situations during fog dispersal operations should be agreed upon by sponsor and operator before award of contract. In the case of an airport operation, the target visibility should be at least the minimum

TABLE 2-2. How A Given Water Content, w_l, Could be Associated with Different Horizontal Visibility Values

w_l [a] ($g\ m^{-3}$)	n_d ($\#\ cm^{-3}$)	d (μm)	V [b] (m)
0.02	50	19.7	324
0.02	100	15.65	257
0.02	500	9.2	149
0.02	1,000	7.25	120

[a] w_l $(\Pi/6) * \rho_l * n_d * (d)^3$ where, ρ_l is the density of water, n_d is the number density of droplets with mean volumetric diameter, d, with $\rho_l = 1\ g\ cm^{-3}$.

[b] $V = A / (4\ d^2) \propto d / w_l$, where V is defined as the distance at which a black target can just be detected against a horizon sky with a contrast threshold of 2%, d is the mean droplet diameter, A is given by $3.9/[pi * n_d * K]$, n_d is the droplet concentration, K is the scattering efficiency factor (typically 2 for droplets), and $\propto$ denotes "is proportional to."

established by the responsible aviation authorities as being required for takeoffs and landings. Fog-clearing operations should continue until the required minimum visibility (i.e., per *FAA* guidelines) is obtained, which allows the airplane to safely land.

2.3 SEEDING STRATEGY FOR DISPERSING SUPERCOOLED FOGS

The strategy for clearing supercooled fogs involves increasing fog hydrometeor sizes to a point where they are large enough to precipitate. Thus, one approach to supercooled fog dispersal could be to modify the strategies for precipitation enhancement in stratus clouds (e.g., Schaefer, 1946a, b; 1948; 1953; ASCE/EWRI Precipitation Enhancement Standard, 42-04, ASCE, 2004). The details of the operational implementation of this strategy will depend on the wind field, air temperature, fog type, fog thickness, number of cloud layers above the fog, and constraints imposed by the operational area, and possibly by the sponsors.

The goal of most supercooled fog dispersal projects can be obtained by creating and placing, within the fog, sufficient ice particles that will ultimately fall to the ground faster than supercooled fog droplets form over, or advect into, an area. The ice particles, or snowflakes, are most commonly created by placing dry ice or liquefied propane into the supercooled fog, although other *glaciogenic seeding* agents may be used.

Once formed, the ice crystals then grow at the expense of the surrounding supercooled water droplets. The idea is to have just enough crystals form throughout the entire cloud area such that the crystals transform surrounding supercooled water and vapor into ice crystals large enough to fall to the ground (Figure 2-1). It is worth noting a common rule of thumb for initiating the ice crystal precipitation process that says, "*. . . at least one ice crystal per liter of supercooled cloud is required to initiate the precipitation process. . . .*"

Maximum seeding effects occur in supercooled fogs that have natural ice crystal concentrations of less than one per liter ($<1\ L^{-1}$). This allows the introduced ice-forming agents to produce ice crystals large enough to precipitate. In the case of a typical *radiation fog*, it is suggested that the introduced seeding agent produce 10 to 50 ice crystals per liter of seeded fog volume (Schaefer, 1951). There is rarely a need to use more than 50 ice crystals per liter of seeded volume. The desirable concentrations needed to nucleate the maximum amount of water to optimize the fog dispersal activity depend on the air temperature, fog-forming mechanism, fog microstructure, location and timing of application, and time available for clearing. The effect of temperature on ice-forming *nuclei* activation is shown in Table 2-1, but temperature may also play a role in the duration of fog (e.g., George, 1951). Natural materials commonly act as ice-forming agents at temperatures between $-8°C$ and $-39°C$. Thus, supercooled fogs with such air temperatures might not react well to seeding operations if there are already sufficient natural *nuclei* present. There are typically insufficient numbers of natural ice nucleants capable of activating the ice phase to disperse the fog in air with temperatures warmer than $-14°C$. It is imperative to remember that the latter is very dependent on localized conditions.

Vickers and Church (1966) conducted some experiments in supercooled clouds to determine the seeding rate that achieved optimum clearing using dry ice pellets. Clouds had thicknesses of approximately 370 m with ambient temperatures of $-3°C$ or colder. Positive results were obtained by applying 1.7 (± 0.1) kg per kilometer of 1 cm cubic pellets with 0.6 $g\ cm^{-3}$ density. Cloud response doubled when they seeded at a rate of 0.85 kg per kilometer of 1 cm cubic pellets with 0.6 $g\ cm^{-3}$ density, all other conditions being the same.

In contrast, clouds failed to respond when they used 0.4 kg per kilometer of 1 cm cubic pellets with 0.6 $g\ cm^{-3}$ density. Mobile dry ice and silver iodide delivery systems have also been used to disperse fogs whose thickness was less than 700 feet.

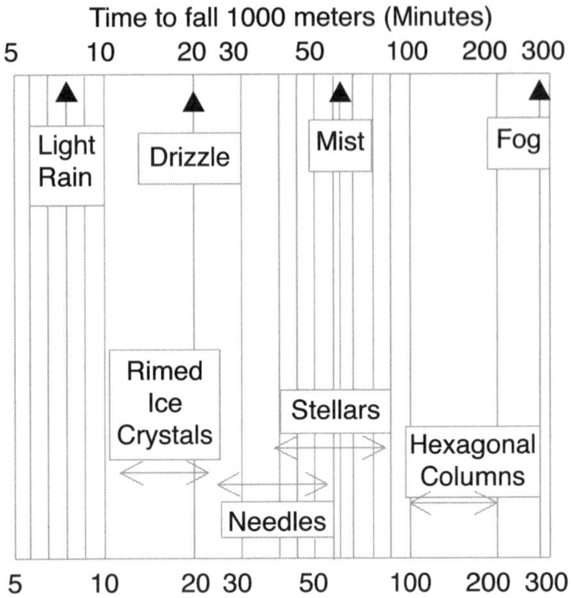

FIGURE 2-1. Size versus fall velocity for various hydrometeors. Stellars are "star"- shaped hexagonal ice crystals. (Adapted from Schaefer and Day, 1981.)

3.0 THE DESIGN OF SUPERCOOLED FOG DISPERSAL PROJECTS

Once supercooled fogs are recognized by a sponsoring entity as causing a problem and a decision is made to sponsor a fog dispersal effort, the next step is to determine the scope of a project to disperse the fog and produce a suitable project design. A suitable project design document includes the project scope; delivery methods; seeding agent selection; targeting and delivery considerations; experience and training; seeding suspension criteria; and legal, environmental, and social considerations. The following is an example of a suitable fog dispersal project design.

3.1 PROJECT SCOPE

The fog dispersal project scope is generally determined by the primary goal of the project, e.g., to clear fog at a specific airport. This immediately defines the geographic area of interest, and a detailed analysis of all relevant information then follows. The analysis should include the geography and topography of the area, the physical characteristics of the fogs themselves, and the weather conditions that typically prevail when the problematic fogs form and persist. Examination of these factors in turn leads to consideration of possible targeting and delivery methods, which culminates in the development of operational procedures and an evaluation plan (Section 5).

3.1.1 Basic project area concepts

Fog dispersal operations do not require the extensive *project* area designs used in precipitation enhancement (ASCE/EWRI Standard, 42-04, ASCE, 2004) and hail suppression operations (EWRI/ASCE Standard, 39-03, ASCE, 2003). Nonetheless, the following briefly describes the relevant concepts of a fog dispersal operation using the terms employed in precipitation augmentation and hail suppression operations as shown in Figure 3-1. The actual orientation and shape of each zone shown in Figure 3-1 should be a function of the target area shape, wind speed, and direction rose during supercooled fog events, topography, and local conditions. For example, the operational area would approach the size of the target area as the wind speed became calm.

1. **Target area**—The target area (solid line 5-sided shape) is the area where the objective is to clear the fog or to increase the visibility to a distance that is suitable for the sponsor to conduct their restricted operations in that area. In the case of an airport, the target area commonly encompasses the runway plus 3 to 6 km (2–4 miles) on each end. Equipment and instrumentation may be deployed within this area to gather data, monitor the operation, or seed *radiation fogs*, since these are typically accompanied by relatively calm winds. Fogs accompanied by winds, such as advection fogs, could require data gathering, monitoring, or seeding that extends from the area covered for the *radiation fog* to upwind of the target area, depending on the meteorological conditions and other project scope considerations.
2. **Operational area**—The operational area (dotted oval) contains the target area and any surrounding upwind areas employed for treatment in order that the effect from seeding covers the target area, as well as for evaluating the project (excluding the control area as long as these areas have not had seeding conducted in them).
3. **Control area(s)**—Control areas (solid ovals) may not be necessary for fog dispersal projects, especially since results of seeding are readily visible upon application. Nonetheless, control areas are still a good idea in fog dispersal projects for documenting the naturally changing physical characteristics of the fog being dispersed.

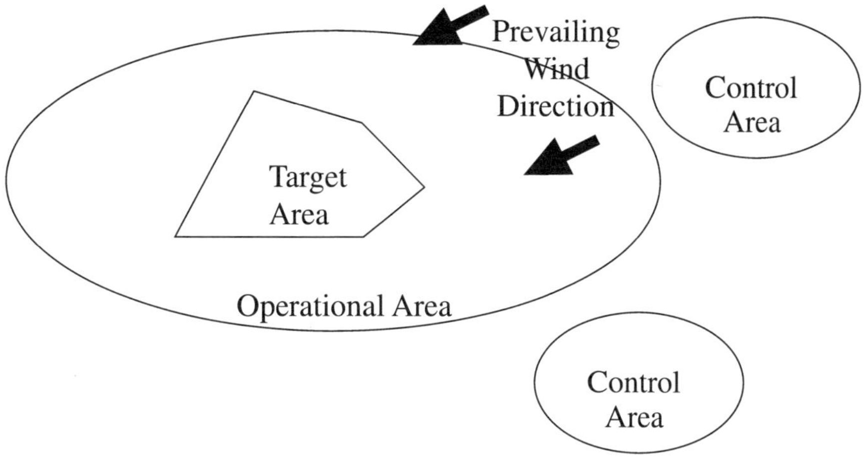

FIGURE 3-1. Terms related to project area and operations for advection fogs.

4. **Project area—**The project area is the area containing all of the equipment used for operations and evaluation or assessment by the contractors and/or sponsors.

3.1.2 Initial design considerations

A project plan should consider:

- where fog clearing is needed
- when during the year (i.e., month(s), season) the project should be operated
- the meteorology (climatology) for this period
- the characteristics of the fogs to be treated
- the treatment method to be utilized
- how the operations will be conducted
- the project decision-making infrastructure
- how project efficacy will be evaluated
- applicable cloud seeding regulations, if any
- if the project is at the airport, that airport's security policies and requirements
- possible suspension criteria

It is highly recommended that all options be considered and evaluated from an economic viewpoint before a final decision is reached to implement a project.

3.2 DELIVERY METHODS

The placement of seeding agents in fog is important for effective fog dispersal. One must decide where, when, how, and how much seeding material should be placed in the fog to maximize the longevity of its dispersal. Important considerations include:

- Where, relative to fog top, the placement of seeding material would yield the desired operational result.
- Time factors, including how quickly a desired level of effect is needed, how long the effect needs to be maintained, and a reasonable estimate of the duration of typical events.

The answers depend on fog depth, air temperature, fog type, wind field (both speed and direction), the seeding agent chosen, and the desires of the sponsor and/or stakeholders. Fog dispersal involves intervening in the microphysical and/or dynamic development of typically nonconvective clouds in order to initiate their precipitation process. Consequently, some of the delivery methods and seeding agents discussed in this section may be similar to those in precipitation augmentation manuals (Griffith et al., 1995; ASCE/EWRI Standard 42-04, ASCE, 2004).

A number of appropriate delivery choices are available for both airborne and ground-based delivery systems. The equipment necessary for a suitable stationary ground-based delivery system can be considerably more expensive than for aircraft or mobile ground-based seeding. This is especially true early in a project's history because there are generally multiple, permanent devices that need to be installed. Mobile ground-based supercooled fog dispersal systems are very cost-effective. Ground-based systems are often very effective when used to disperse rather shallow fogs (i.e., less than 700 m in depth), whereas they are less effective in dispersing fogs greater than 700 m thick. However, it should be noted that it would seldom, if ever, be necessary to totally disperse such a deep fog through its entire depth. Dennis (1980) provides additional details in support of the 700 m limit.

Fog dispersal operations are conducted most efficiently when natural processes, such as turbulent diffusion, help spread the seeding material. The process of turbulent diffusion spreads the seeding agent or ice crystals in three dimensions throughout the fog. This process is sometimes limited by the fact that fog forms in stable air. If an aircraft drops dry ice forming curtains of ice crystals, diffusion has to spread the ice crystals in only one direction. For example, a typical airborne supercooled fog clearing operation at an airport has its aircraft dispense approximately 1 or 2 kg km^{-1} of dry ice sized to just reach the ground when released from 500 m above the ground. Each aircraft pass dispenses dry ice parallel to, and 45 to 60 minutes upwind of, the runway to be cleared. This creates an ice crystal curtain that drifts downwind while spreading due to turbulent diffusion. Each pass is repeated frequently enough so that the resulting ice curtains are about 1 km apart. This helps to minimize the refilling of air over the runway due to turbulent mixing with untreated air. Assuming a line contains an infinite number of point sources, with a typical stability class of A or B (Pasquill, 1962), and a wind speed between 1 and 2 m s^{-1}, then the vertical dispersion (or depth) of the seed plume would reach about 700 m at a location 3 km downwind. This same natural process is of limited help during calm conditions. These situations require the aircraft to spread sufficient seeding material to treat the target area and to repeat this treatment as often as needed.

A point seeding source on the ground requires turbulent diffusion to spread the ice crystals or the nuclei themselves both vertically and horizontally. Mobile, ground-based delivery systems, dispensing either dry ice or silver iodide, are employed most effectively when the fog thickness is shallow and, in the case of silver iodide, when air temperatures are below −4°C. Most supercooled fogs are often only slightly below freezing temperatures (i.e., approximately −2 to −5°C). This means the best seeding agents are those that operate effectively at these slightly supercooled temperatures (i.e., approximately −2 to −5°C). Silver iodide is seldom used in supercooled fog suppression operations since silver iodide's activation threshold is near −4°C with increasing activity at colder temperatures.

3.2.1 Airborne projects

Assuming the fog is thick enough to warrant the use of aircraft, consideration should be given to the flight path and speed of the aircraft during seeding, the seeding agent, and what equipment to use.

The typical airport clearing flight path encompasses the runway plus up to 3 to 6 km (2–4 miles) on each end (Figure 3-2), depending on the meteorological conditions, particularly the wind field, and air traffic control. These dimensions will vary depending on the spatial extent of the actual fog, runway size, and program scope, as well as criteria determined by the Federal Aviation Administration (FAA) or similar authorities. For example, at FAA-controlled airports, a category I landing system requires a minimum visibility of 732 m and a decision height of 61 m. A category II landing system requires a minimum visibility of 305 m and a decision height of about 31 m. Category III landing systems do not have a decision height specification but require 213 m for subcategory "a," 46 m for subcategory "b," and 0 m for subcategory "c" (Silverman and Weinstein, 1974). Other possible flight paths such as the racetrack pattern (Figure 3-3, **a**) and the raster pattern (Figure 3-3, **b**) can be used. The race-

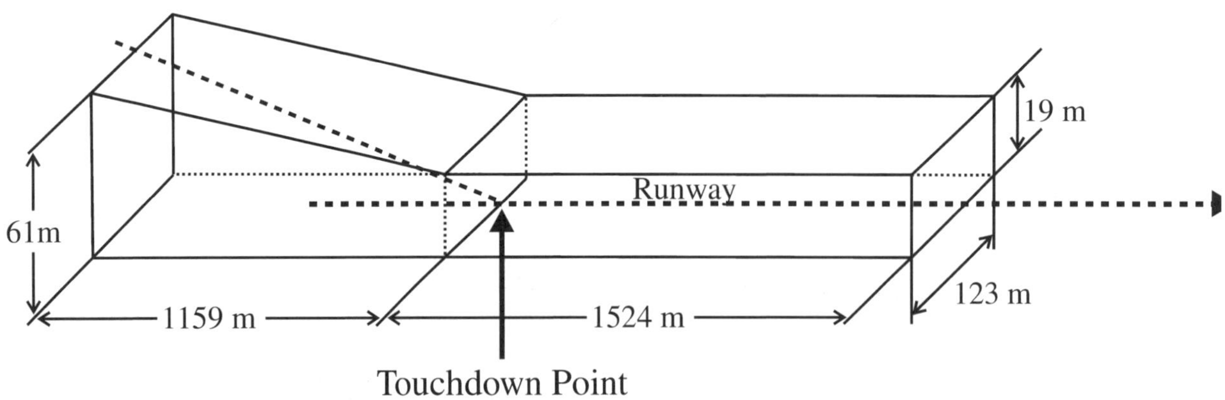

FIGURE 3-2. Air space typically cleared during airport fog dispersal operations. The dimensions are derived from the typical dimensions (in meters) provided by Silverman and Weinstein (1974). The thick dashed arrow represents the path of the airplane as it approaches, lands, and travels along the runway as well as the direction of departures.

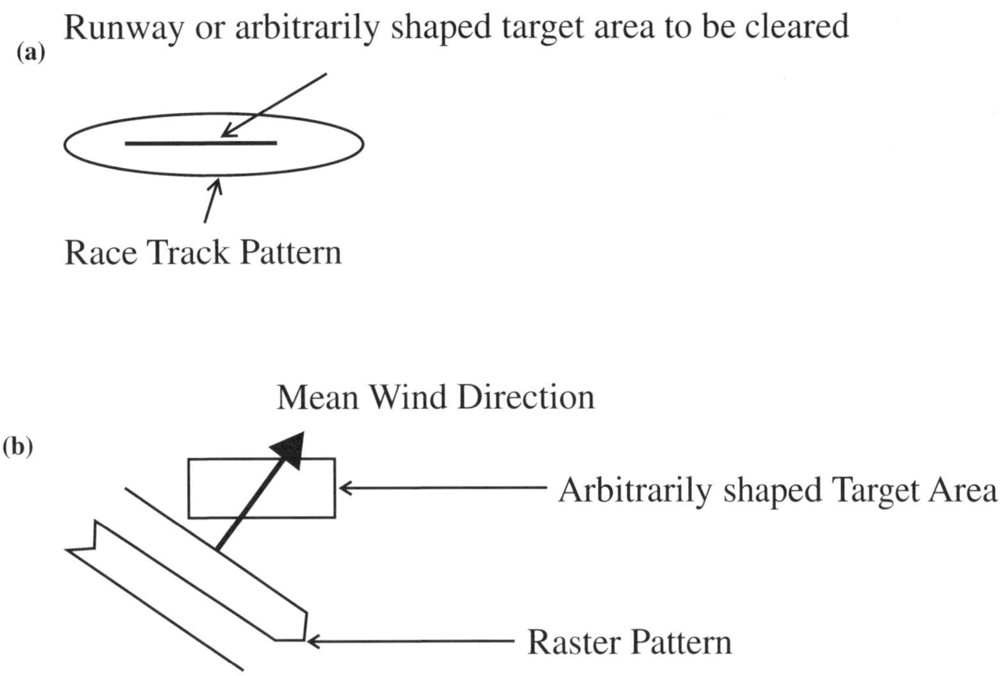

FIGURE 3-3. Two possible flight paths for clearing supercooled fog decks: a, The race track pattern; b, the raster pattern.

track pattern does not typically extend the entire size of the airport, but extends just beyond the runway dimensions. The racetrack pattern works best in radiation fog situations (e.g., no wind flows). The spacing between the seed lines of the raster pattern is typically 1 km (Dennis, 1980).

Airborne fog dispersal programs favor dry ice or liquefied propane, although silver iodide generators or burn-in-place flares can be used if air temperatures are cold enough. We *do not recommend* the *use* of *ejectable flares* when *dispersing fogs with thicknesses under 1 km*. This is founded on common sense, but such use generally violates safety and ceiling height regulations administered by the FAA. The FAA can provide waivers for altitude restrictions during airborne fog dispersal operations. Three reasons for the preferred use of dry ice to disperse supercooled fog are (1) the warmer ice nucleation threshold (i.e., Table 2-2), (2) dry ice or liquefied gases cover a greater vertical extent than the seeding agent from the silver iodide generators or burn-in-place flares, and (3) operational flexibility. Dry ice pellets have a fixed fall distance that depends on the air temperature, wind field, and their size (Vickers and Church, 1966). They are slow to sublimate, and typical, commercially available sizes easily fall 1 km. Thus, if the seeded fogs are shallower than 1 km and dry ice is used to disperse them, the dry ice pellet sizes should be reduced in a roughly inverse linear manner. That is, a half commercial size dry ice pellet will fall approximately 500 m. The dry ice must be loaded and handled with some care, but it requires minimal preparation compared to some alternatives. Details on some of the equipment used in seeding operations can be found in the ASCE Manual of Professional Practice No. 81 (Griffith et al., 1995).

Fog dispersal operations should be conducted using a twin engine aircraft certified for flight in known icing conditions. The aircraft must have an experienced pilot in addition to adequate navigation equipment, and it must equipped with the appropriate seeding equipment to conduct the operation from within or above the fog deck. Two engines will help ensure safety. A single engine aircraft is not advisable for safety reasons: If it loses its engine, a safe landing could be difficult. Nonetheless, aircraft types should be determined after careful consideration of all operational factors including the maximum fog thickness in the area chosen for seeding, the areal extent of the operational area, and the availability of suitable alternate airports.

3.2.2 Ground-based projects

Liquefied propane, nitrogen, or carbon dioxide ground generating systems, or mobile dry ice platforms are particularly attractive for airports located in mountainous regions of the western United States where airborne seeding methods might not be practical. Mobile silver iodide delivery systems have also been used to disperse shallow fogs (thicknesses less than 700 m) when typical air temperatures were at or below −5°C.

The number and position of the generators within a network should be determined by considering at least the following factors:

- The characteristics of the emission, the type of generators, and the seeding agent.
- The background concentration of naturally occurring ice-forming nuclei (IFN) in the target area.
- The meteorological characteristics during conditions conducive to the formation and/or advection of fog in/into the target area and how they relate to dissipation efforts.

Ground dispensers should be situated to ensure that the wind and any natural turbulence will carry the seeding material in the desired directions and assist in its dispersion on its way to and across the target area. Generator networks can be designed and used with the aid of a computerized dispersion model. That will allow visualizing the seeding agent trajectory to optimize operations. This technique can also be used if the ground platforms are mobile.

The liquid propane dispenser used by the Paris Airport Authority at Orly Airport, France, since 1964, dispensed up to 10 gallons of propane per hour at an operating cost of $20/hour in 1970. The Orly Airport had 60 such dispensers—four near the airport for calm winds, and the rest placed at various distances upwind along the most common wind directions. The Air Force Weather Agency at Fairchild Air Force Base, Spokane, Washington, also had a similar system. Silverman and Weinstein (1974) provide some additional details.

3.3 SEEDING AGENT SELECTION

The materials placed within a targeted fog are known as fog seeding agents, and they may be dispensed via ground or airborne delivery systems. The two most common supercooled fog dispersal seeding agents are dry ice and liquefied glaciogenics, such as propane. Silver iodide is not commonly used to disperse supercooled fog even though it is frequently used in the conduct of precipitation enhancement programs. The differences between the manner in which dry ice or liquefied glaciogenics create ice crystals are discussed in the following subsections. A variety of liquefied glaciogenics (e.g., propane, carbon dioxide, nitrogen, etc.) are capable of creating ice crystals in supercooled fog-clearing operations.

Pyrotechnic formulations are available that produce effective hygroscopic nuclei, but their effective use in supercooled fog dispersal operations is unproven. The hygroscopic flare offers the benefit of relatively low production costs. When used with an airborne platform, there are essentially no corrosion problems and a significantly reduced need for expensive pressurized aircraft. Hygroscopic agents become liquid by "absorbing" water vapor from the air (i.e., deliquesce) at relative humidity values significantly less than 100%. Salts within pyrotechnic flares are generally used for enhancing the precipitation process in warm clouds. Finnegan and Pitter (1997) discuss a mechanism that provides a basis for why such formulations might be effective in supercooled fog dispersal.

3.3.1 Dry ice, liquid propane, and liquid nitrogen

Dry ice (Figure 3-4) rapidly transforms vapor into ice crystals (Schaefer, 1946a, b; Vonnegut, 1981), which then grow at the expense of the surrounding supercooled droplets. Any transformation of preexisting fog droplets directly into the ice phase is a secondary effect compared to the conversion of water vapor into ice crystals (Vonnegut, 1981). Dry ice does not work consistently at air temperatures warmer than about −1°C to −2°C. Dry ice should be used when air temperatures accompanying the fog being treated are between −2°C and about −10°C. Dry ice is normally dispensed from aircraft. Dry ice can be obtained in pelletized form as pictured in Figure 3-4. This form is more convenient to use than obtaining dry ice in blocks or sheets and then breaking or sawing it into smaller sizes either on the ground or crushing it onboard the seeding aircraft. Dry ice is difficult to store because it continuously sublimates. Consequently, supplies of dry ice must be continually replenished during periods when supercooled fog is forecast to occur.

Dry ice produces on the order of 10^{11} to 10^{12} ice crystals per gram of dry ice dispensed, and this production may be dependent on temperature in a range from −1°C to −11°C (Holroyd et al., 1978; Horn et al., 1983).

Dry ice has the advantage of being a natural substance (i.e., frozen carbon dioxide), being very effective at temperatures only slightly colder than 0°C,

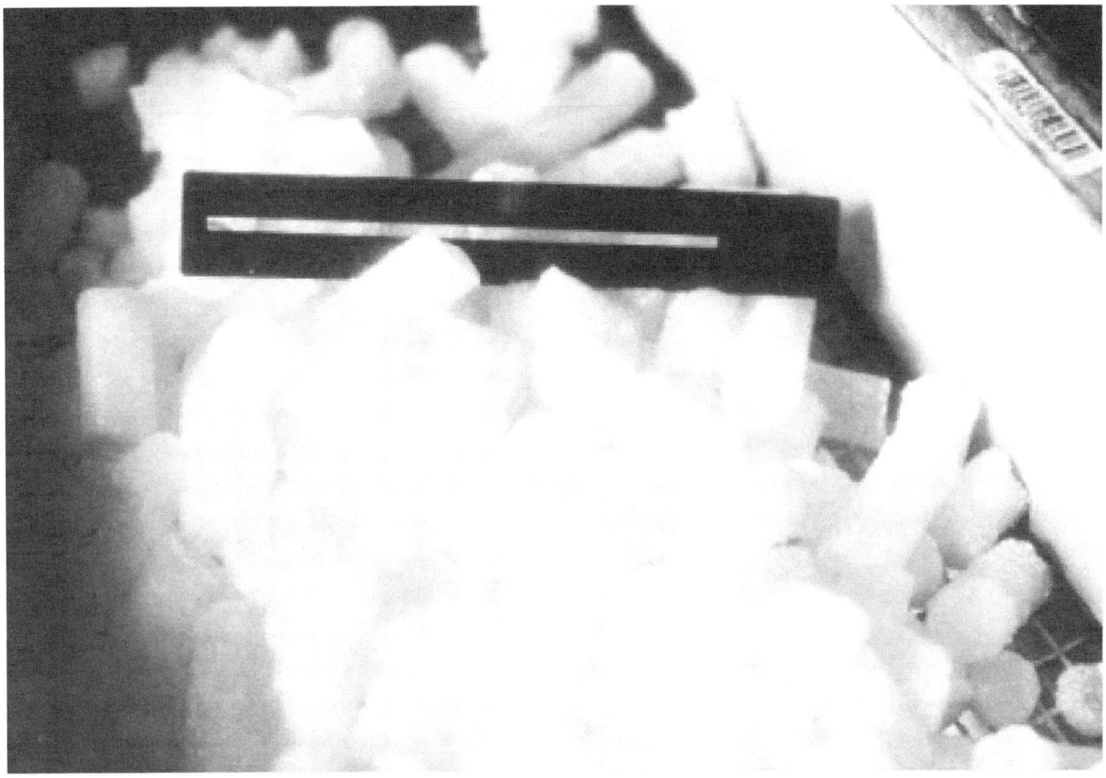

FIGURE 3-4. Dry ice pellets, in protruded form, used in airborne dry ice fog dispersal operations. Length of ruler is 6 inches.

generating ice crystals nearly instantaneously, producing ice crystal numbers almost independent of air temperature, and being relatively inexpensive.

Liquid carbon dioxide, liquid propane (Figure 3-5), and liquid nitrogen produce ice crystals in a small cone-shaped zone of extremely low temperatures that is produced when one of these liquefied, pressurized gases is vented into the air. They produce nearly the same number of ice crystals per gram as dry ice (Kumai, 1982), but can also be dispensed from the ground. Liquid propane is not dispensed from an aircraft because it is flammable. Liquid nitrogen has been dispensed from aircraft. The United States Air Force (USAF) has used ground-based liquid propane to clear supercooled fog from military airports for over 20 years (Griffith et al., 1995). The possibility of flammability does exist with the liquid propane dispenser systems, and as a result these systems should be completely surrounded by a fence. The propane odor could be a nuisance factor and a possible concern to the public. "Unstenched" propane is available and is used by the military, e.g., Fairchild Air Force Base, Spokane, Washington, but it is not commercially available.

3.3.2 Quality control of seeding agents

Program planners will normally purchase their own seeding agents from commercial suppliers. The following factors should be considered:

- The quality of the seeding agents used.
- The ingredients (active and inactive with respect to ice nucleation) of all seeding agents, including flares, should be specified along with their quality and concentration.
- Material safety data sheets (MSDS) must accompany all materials, whether in raw form or ready to use. Dry ice pellets must be of consistent sizes, and transported and stored in containers that minimize sublimation and limit the development of water ice within the dry ice.

3.4 TARGETING AND DELIVERY CONSIDERATIONS

The consistent proper placement of seeding agent(s) in both time and space is defined as targeting and is absolutely essential to produce the desired

FIGURE 3-5. Example of propane dispenser. (Photo courtesy of J. Heimbach.)

results. Targeting is one of the more difficult tasks in any cloud seeding program. Fog dispersal operations may have significant targeting concerns, especially when turbulence and vertical wind shear are present. Turbulence and vertical wind shear are a concern when seeding advection fog. Thus, the choice of seeding agent may determine the choice of delivery system. Keep in mind that the volume of air that one is attempting to keep clear is on the order of 10^7 m^3.

Turbulence and vertical wind shear help mix seeding agents into and within treated fogs. Most fogs form in stable atmospheres, so vertical transport of seeding agents released from the surface is not aided by convective mixing. Turbulent mixing near the surface increases the depth (height) to which the seeding agent may be mixed. Hence, shallow fogs are best targeted by ground-based seeding methods. However, since the upper portion of the fog remains unaffected, solar radiation that might help keep the fog from re-forming cannot reach the ground. Deeper fogs can be seeded from above to ensure that the full depth of the fog bank is treated. Ground releases in deeper fogs may be effective, however, since it may not be required to clear the entire fog, but to increase the visibilities in the landing zone of the airport above some minimum values as discussed in Section 3.2. If the upper portions of the deeper fog decks are effectively targeted via aircraft seeding, the resulting ice crystals may "sweep out" unfrozen fog droplets below as they fall to the surface, effectively clearing the lower portion of the fog as well.

Predictions of how seeded volumes of foggy air will be advected over the target area are also important, since this will determine where treatment must take place in order for it to be effective. Vertical wind shear affects the trajectory of ground-based and airborne-released seeding agents as well as any precipitation that may result from seeding. For example, consider a 300 m thick supercooled fog will be dispersed over an airport runway using an airborne dry ice delivery system. The aircraft drops a curtain of ice crystals with 5×10^8 crystals m^{-2}, seeding at a rate of 1 kg km^{-1} perpendicular to a mean wind field of 1 m s^{-1} and extending 1 to 2 km beyond each edge of the runway. The aircraft travels at 50 m s^{-1}, and the runway is approximately 2,700 m long. This curtain is predicted to reach the airport runway in 3,000 s. Its width will spread (in both directions) at an approximate rate of 1 to 2 m s^{-1} and will become approximately 1 km wide over the runway, assuming reasonable values of turbulence (Dennis, 1980). The ice crystal concentration will decrease to about 100 per liter due to turbulent spreading, aggregation, and fallout just before reaching the runway. A few of these ice crystals will grow to snowflake size and fall to the ground in 10 to 20 minutes after the aircraft drop, while the smaller ice crystals continue growing until they too fall out. The ice crystal concentrations should be near 2,000 per liter, slightly higher at the center, within 10 minutes of the drop, and practically all of the supercooled water would be removed within the extent of the curtain due to evaporation and accretion growth (Dennis, 1980). Returning to the example, as the seed plane turns around to head back to its origin point, clearing should be evident. However, seeding cannot cease, because turbulent mixing with untreated air could cause the cleared fog to fill in within 50 to 100 minutes after the dry ice drop. Thus, in order to disperse the fog over the runway, the seeding lines should be dropped continuously about 50 minutes upwind of the runway and repeated frequently enough so that the resulting curtain centerlines are about 1 km apart. The frequency depends on the mean wind speed in the fog deck, how fast the seed aircraft travels while dropping the dry ice, the visibility requirement at the runway, and how long the visibility requirement needs to be maintained.

An aircraft drops 2 to 4 kg km^{-1} of dry ice on seeding passes that take place about 45 to 60 minutes upwind of the target area in a pattern that usually consists of 5 to 30 parallel lines, spaced between 800 and 2,400 m apart during a typical supercooled fog-clearing operation. These passes typically range from 8 to 10 km long. A mobile liquefied glaciogenic ground-based delivery system can accomplish similar results by driving back and forth along a road parallel to the runway to be cleared (this assumes that there are roads available at the right distances upwind that will allow this type of seeding to be done). See Silverman and Weinstein (1974) and Dennis (1980) for additional details. A typical fog dispersal operation could require about 500 kg of dry ice, cost approximately $80,000 (winter 1969–1970 U.S. dollars), and save over $900,000 in airline operating expenses (Silverman and Weinstein, 1974).

3.5 EXPERIENCE AND TRAINING

Operational fog dispersal personnel require special knowledge and skills. National, state, and local regulations should be consulted to determine whether personnel in control of operations are required to be certified *Weather Modification Operators* or *Managers*, as defined by such organizations as the Weather Modification Association (WMA, 2003). Some states in the United States have such requirements. If the particular government agency does require certification, the regulatory entity should be contacted for the specifics of the requirements. The *Weather Modification Association* may be able to furnish information on such regulations. Some of the required skills are discussed next. Certification demonstrates a proven level of competence and ensures that the field personnel have the special knowledge and skills required to perform successful fog dispersal operations.

3.6 SEEDING SUSPENSION CRITERIA

Suspension criteria are developed to protect the physical and economic well-being of the *project* area or to prevent damage to the environment. Their fundamental scope is to reassure the public that the weather modification project is properly run in such a way that these criteria spell out adequate safeguards designed to avoid, insofar as possible, damage to the public health, safety, or welfare, or to the environment, and does not substantially interfere with economic or social activities.

Fog-clearing activities should be suspended whenever conditions develop in which continued operations could potentially be a threat to the health, safety, or welfare of area residents. Perception is very important. In the case of supercooled fog dispersal, reasons for suspension might include unsafe flight conditions and/or roadways being used by mobile seeding platforms, a weather-related accident within or immediately downwind of the seeding area, and generally hazardous weather such as freezing precipitation. Fog dispersal operations pose no significant environmental threat. See Section 3.7 for additional details.

3.7 LEGAL, ENVIRONMENTAL, AND SOCIAL CONSIDERATIONS

Liabilities have been limited in weather modification and are virtually nonexistent with respect to supercooled fog dispersal projects. The city of Boise, Idaho, was sued in the early 1990s because there was a car accident in the vicinity of the local airport following an airborne fog dispersal seeding event. As a result of this lawsuit, they now perform such operations via a mobile ground dry ice delivery system in association with public advertisements. Fog-associated automobile accidents cost insurance companies in excess of $300 million/yr (Silverman and Weinstein, 1974). Thus, a key societal benefit of fog dispersal projects could be the prevention of multicar pileups on fog-shrouded highways.

The concessions at major airports suffer during foggy periods from reduced sales of their food, gifts, and rental cars, down to parking to ultimately local motels, restaurants, and even the airlines. Northwest Airlines Bismarck, North Dakota, officials indicated that rerouting planes due to fog could cost an amount equivalent to that for fuel and engine time, as well as in increased ground crew efforts/time. Consider the cost of flying back to the airport of origination, which in this case would mean a return to Minneapolis. The entire flight must be repeated or the passengers bused to their destinations. Sometimes flights are unable to depart in fogs; in which case not only is the revenue for that flight lost, but also the aircraft, which was scheduled for multiple flights on that day, becomes unavailable for all other flights.

The societal benefit of seeing the sun during otherwise dreary periods cannot be emphasized enough. For example, it could be a personal advantage to those suffering from "seasonal affective disorder" (SAD). More dependable airport operations would result from fog dispersal operations. The increased revenue compared to periods when facilities had to otherwise shut down on regular bases should also have a positive but indirect effect on the social well-being of the affected population. If the increased revenue is large enough, then perhaps local schools and the community will benefit. Fog dispersal projects do not pose any threat to the environment (e.g., Douglas, 1968; Cooper and Jolly, 1969; Cooper, 1977; Klein, 1978a, b).

4.0 SUPERCOOLED FOG DISPERSAL OPERATIONS

This section contains some guidelines for operational decision making and project infrastructure. The scope, objectives, methodology, decision making, and communications infrastructure of the project should be clearly defined and provided to all operational personnel as part of a project operations manual. Development of the operations manual requires advanced thought and study of all aspects of the project. A daylong ground school session on fog dispersal operations would be a good venue for this information. The general requirements for the conduct of supercooled fog dispersal operations, including notification of seeding intent to authorities, selection of seeding agent(s), delivery systems, support equipment, targeting, opportunity recognition, and personnel requirements are discussed in this section.

4.1 OPERATIONS MANUAL

An operations manual should be constructed and made available to all project personnel. The operations manual should provide clear definitions of all terms used in conducting operations, describe the types and functions of all information-gathering systems employed and all actions to be undertaken, define all phases of operations, and state criteria for initiation and termination of each phase. Procedures for recording and reporting operational information should be given, particularly to comply with any weather modification laws. The circumstances under which seeding will be suspended should also be clearly defined so that prompt decisions can be made when circumstances warrant. Those establishing new projects are advised to obtain copies of manuals used on other projects. Certified managers or operators from commercial cloud seeding firms can usually answer any questions raised while creating a new program-specific manual.

4.2 PERSONNEL REQUIREMENTS

The typical fog dispersal ground-based operation requires one full-time person to maintain and operate all equipment and make observations, assuming a single, mobile, ground-based seeding system is used. This person should be a skilled forecaster and be technical enough to handle basic electronics technician tasks (two-way radio communications, computing equipment) and maintenance of ground-based silver iodide generators, liquid propane generators, and mobile platform systems. National, state, and local laws should be consulted to determine whether personnel in control of operations require certification. The sponsor should be aware that the Weather Modification Association is proactive in ensuring that all operators and managers of a fog dispersal project (or any weather modification type of project) possess the knowledge, experience, character, and good morals as expressed in the WMA professional code of ethics (WMA, 2004). Please note that states do not necessarily require this. These include, but are not necessarily limited to, (1) compliance with all laws and regulations of the federal, state, and local governmental units having jurisdiction in the areas where the projects are conducted; and (2) nonparticipation in any activities that can reasonably be expected to be detrimental to the general public interest or to contribute to undue hardship in operational areas.

4.2.1 Meteorological Staff

A dedicated project forecaster is beneficial since such a person has more at stake than might a NWS forecaster. Unforecasted fogs may result in missed opportunities and unnecessary economic losses. The full-time person assigned to the fog dispersal program would preferably be a certified weather modification manager or operator and/or a meteorologist. The WMA has established processes for the certification of *weather modification* operators and managers (WMA, 2003).

4.2.2 Support Personnel

Part-time support of an experienced technician should be arranged if the full-time person is unable to handle all these tasks. Thus, a fog dispersal operation requires a minimum of one full-time meteorologist, perhaps with part-time support of a technician. If the project involves airborne treatment, a pilot and possibly a copilot will also be needed.

4.2.3 Cloud Treatment Pilots

In the event of aircraft being involved in a project, an experienced pilot will need to be able to maintain regular radio contact with the other project personnel and air traffic control, etc. If aircraft are used, an analysis could be performed (based on climatological data) to determine the number of aircraft required at any time (e.g., how many fogs occur in the target area, on any day during the proposed operational period, are there consecutive fog days, and what are their thicknesses). This information should allow one to determine whether the use of aircraft would be wise, and if so the questions of where to locate the planes, and of pilot training and experience present themselves, among others. There must be an alternate airport plan in case the seed aircraft is fogged in. Airborne operations will also require an airplane mechanic and an electronics technician for all equipment maintenance on the aircraft in addition to the other electronics needs for the fog dispersal operation.

4.2.4 Direction of Operations

In the unlikely event of larger projects, there might be separate people to act as the operations director, meteorologist, electronics technician, pilot mechanic, and pilot (if aircraft are used). A person experienced in fog dispersal operations should direct operations from the operations center. The operations director must have access to important decision-making information while at the operations center. Such decisions typically concern the initiation of liquefied gas releases, the launching of aircraft, and the termination of operations.

4.3 OPERATIONAL DECISION MAKING

The National Weather Service (NWS, United States), Meteorological Service of Canada, and other analogous organizations in other countries provide generalized forecasts about expected foggy periods, and these should be made available to the operators of fog dispersal projects in the respective areas. In particular, it is important to be aware of fog advisories issued by the official government weather agency in the region. Maintaining awareness of fog advisories will enable the operator to maximize the number of successful fog dispersal events. The person running the project must determine the meteorological conditions near the target area and make a decision to seed or not to seed, what delivery method and seeding agent to use, and how much seeding material should be dispensed.

The forecasting of fog is locale-specific, and the forecasters of fog dispersal projects (or any weather modification effort) should at least visit and listen to the elders of the area in which their operations will exist, whether they are stationed in the field for the particular project period or not. An on-site visit will provide the forecaster(s) with a sense of the surrounding topographical features and some insights of the local influences on the weather in the area for which they are forecasting.

Fog climatologies for the region should be used when developing the operations plan. Most projects begin and end each daily operational period with a weather briefing, wherein project staff is provided with a forecast of the likelihood of treatable fog during the upcoming 24-hour period. The primary purpose of the briefing is to heighten the likelihood of timely opportunity recognition. Generally, in fog dispersal projects the "prime time" for operations is early morning through pre-noon, with a secondary maximum from late evening through just before midnight. Fog may occur any time of the day, so there are no hours during which operations cannot occur. If fog already exists at the time of the scheduled briefing, operations will be under way. However, the forecast is still given so that personnel will have some idea as to how long they are expected to continue operations. When it is determined that the conditions are favorable for fog formation, the forecaster must alert the operations crew, as well as the sponsor and, if applicable, the official(s) in charge of airport operations.

After the weather briefing (if operations are not under way), duty assignments for the day are carried out. Otherwise personnel are on standby and furnished with pagers or cell phones for notification of pending operations. Once operations begin, staff member(s) is (are) engaged accordingly. If no suitable fog develops, the operational status might be downgraded. The whole cycle repeats itself every day.

4.4 COMMUNICATIONS

The project personnel must communicate freely and effectively. Others with whom regular and reliable communication is necessary include project sponsor(s), the local weather service office (if such an office exists), and people within the target area. For fog dispersal projects at airports, this would include the airport authorities and possibly the control tower. If airborne operations are conducted near an airport, ongoing seeding operations might also be noted on the airport's prerecorded field information radio broadcast, now common at most medium and large airports in North America. Pilots approaching the field listen to these broadcasts before contacting the control tower, and so would immediately become aware of the presence of seeding aircraft in and near the fog. Conventional telephones and two-way radios are most commonly used, although cellular telephones are now a viable alternative in most locations. Project personnel may also carry electronic devices that tell the bearer to "phone in" for further information. How to reach somebody is just as important as who to contact in a given situation.

There may be the need to suspend operations because of a perceived public hazard. In such cases, those who need to know must be notified immediately. It is also important that provisions be made to document all operational decisions.

Project planners must consider how the project will be managed and establish a reliable decision-making chain of command. All this must be contained within the operations manual.

4.5 PUBLIC RELATIONS/INFORMATION/ INVOLVEMENT/SAFETY CONSIDERATIONS

The public should be provided with factual information about the goals, objectives, and methodologies of the project. Failure to provide such information could generate undue negative press.

Many communities require hearings before a weather modification project can be undertaken or before a license or permit for the project can be granted by the state licensing and permitting agency. Concerned individuals will generally present themselves at such meetings. Their concerns should be addressed with respect and listened to carefully. The concerns that can be immediately addressed should be. Others may require some additional information gathering, which should be done and the concerned parties informed of the findings. Some objections, such as conflict with religious beliefs, cannot always be addressed in ways that the objector finds acceptable. Concerns about adverse environmental effects from the seeding agents, cost, or making the visibility worse are readily addressed. Several states that have weather modification laws do not require licenses or permits for fog-clearing operations.

Criteria for the suspension of operations during a supercooled fog dispersal project should be delineated in the project operations manual. At least one copy of the operations manual should be onsite and placed in a location that is readily accessible, viewable, and otherwise available to the public. Special weather statements issued by the responsible governmental agencies (e.g., National Weather Service in the United States) and statements from the local authorities may also be incorporated within the suspension criteria.

Public safety is always a concern, even in cases of fog dispersal. Information brochures about weather modification operations, answers to common questions, and the results can be obtained from most long-term projects and from the WMA. These brochures may be helpful in more ways than one, as they may provide ideas for locally produced materials that are specifically tailored to the project at hand.

There are a few hazards associated with fog dispersal operations, besides the obvious hazard during taxi, takeoff, and landings in poor visibility conditions. Flammable and/or explosive chemicals, oxidizers, and solutions may be employed as seeding agents or as ingredients used in the formulation of seeding agents for fog dispersal. These must be stored in a safe area, approved by local officials. Most fire departments also appreciate being made aware of the locations, quantities, and characteristics of such materials being stored, so that in the event of fire, they immediately know what must be dealt with. For example, some chemicals used in seeding formulations, such as silver iodide (AgI), are not combustible in raw form and pose no fire or explosive threat. Oxidizers such as sodium perchlorate will burn vigorously at very high temperatures if ignited and may explode in confined containers. These must be stored only in approved, appropriately labeled containers and in suitable locations. Liquid seeding solutions are primarily acetone-based. Acetone is highly volatile and poses a significant fire risk. It is of low viscosity (splashes easily) and produces potentially toxic vapors. Acetone solutions must be prepared and dispensed only by trained personnel, equipped with rubber gloves and protective eyewear. Eyewash stations should be provided to field sites where the solution is handled, just in case. When handled outdoors, the acetone vapors are very unlikely to become concentrated enough to be considered hazardous. Most local fire departments require the storage of acetone and acetone solutions in "spill-proof" overflow containers.

Another hazard is driving in low visibility areas, and this is especially true of operations involving mobile generators. The most common hazard created by fog dispersal projects is ice-phase precipitation, i.e., snow, which may slicken roadways. Only minimal accumulations generally result, and the resulting improvement in visibility and elimination of the supercooled fog more than offsets this. It may prove helpful to notify the local street or highway department to ensure that sanding of affected roads is conducted if needed.

Dry ice, or solid carbon dioxide, has an equilibrium surface temperature of −78°C. If allowed to come in contact with bare skin, freezing can occur in an extremely short time. Dry ice should therefore be handled, sifted, and loaded into seeding equipment only by those personnel wearing insulated gloves. Project personnel should also be made aware that dry ice continually sublimates at normal atmospheric pressures and temperatures and therefore displaces the oxygen needed to sustain life. Personnel should never transport dry ice within the passenger compartment of any vehicle, and all dry ice dispensing equipment onboard aircraft must be vented to the outside. All project personnel who will be working with seeding agents or in their proximity must be provided with material safety data sheets (MSDS) regarding each of the seeding agents so that they are fully aware of all proper and safe handling and storage requirements, as well as the potential hazards.

5.0 EVALUATION OF SUPERCOOLED FOG DISPERSAL PROJECTS

Project evaluations are conducted to determine how successful the operations were in achieving the desired results. The desired result is most commonly increased visibility in the target area. Supercooled fog dispersal efforts are typically evaluated by line-of-sight visibility. Project evaluations are also important when researching new, more efficient ways to improve upon the present fog dispersal technologies, and techniques. Project sponsors are strongly encouraged to plan some evaluation procedures into their project design. Supercooled fog dispersal is generally free of elaborate evaluation efforts.

Supercooled fog dispersal technology is well established, is easily applied, and usually produces rapid results, reducing the need for an evaluation of such projects. However, evaluations of cloud seeding projects may be required by the project sponsor.

The most direct, and perhaps the best way to evaluate a supercooled fog dispersal project is to continuously monitor the visibility in the control and target areas before, during, and after seeding. The monitoring device could be a video camera, vertically pointing cloud radar (i.e., Ka or Ku band radar), satellite imagery, or some combination of these. The visibility data from the surrounding region may be obtained from National Weather Service (NWS) staffed and automatic weather stations. Even though there are limitations to the values of visibility reported by the National Weather Service, the measurement is still sensitive enough to indicate a positive or negative result, due to the fog dispersal attempt. For example, the NWS station observations will report only a visibility up to 10 miles, even if the actual visibility is greater than that value. Furthermore, the NWS automated weather observing system (ASOS) (e.g., DeFelice, 1998) may report a different visibility from that in your target area for a number of reasons, such as mesoscale variability contributions to visibility. The NWS reports visibility in miles ranging from zero to 10. The start time of the operation should ideally coincide with those periods when the visibility falls below 1 mile for at least 15 minutes; the same is true for operation end time. The amount of seeding material used to dissipate the fog is typically used as part of new research into new, more efficient ways to improve upon present fog dispersal technologies.

The evaluation of supercooled fog dispersal projects may be conducted by directly measuring and analyzing other physical (i.e., not visibility) and chemical properties of the supercooled fog intended to be cleared before, during, and after seeding occur. This kind of evaluation is used when a research project is piggybacked onto the operational project. Although such is rare, it usually involves some or all of the following equipment: onsite radiosonde or vertical balloon profile (e.g., Wetzel et al., 1996), ground-based icing rate meter (Hindman, 1986; Kahan et al., 1995), satellite data (e.g., Welch and Wielicki, 1986; Welch et al., 1986; Ellrod, 1992, 1995; Purdom, 1996; Wetzel et al., 1996; Schemenauer, 1998), and *remote sensing devices* (such as vertically pointing cloud radar, i.e., Ka or Ku band radar, lidar, dual wavelength microwave radiometer). DeFelice (1998) provides general background on remote sensing measurements used for meteorological applications, including details on their operation. Such an effort might also involve a modeling component (e.g., Welch et al., 1986; Bott et al., 1990; Duynkerke, 1991), which requires measurements of standard meteorological weather parameters, droplet size distribution, and optical path, among others.

6.0 GLOSSARY OF ACRONYMS AND TERMS

Italicized print in this section indicates a glossary entry.

Advection. A local change in the properties of a fluid (e.g., temperature, relative humidity) caused by its movement. This movement is usually considered to be in the horizontal direction but may be used in terms of vertical movement.

Advection fog. Advection fogs are produced from the movement of warm, water vapor–laden air over a surface cold enough to induce condensation process. Fogs along the California coast are generally advection fogs. Advection-radiation and upslope fogs are special cases of advection fogs.

Advection-*radiation fog*. Advection fog that subsequently moves over relatively colder ground that continues to cool by radiation.

AgI. See *silver iodide*.

Air mass. A large body of air (hundreds to thousands of square kilometers) that possesses the same temperature and dew point temperature level by level.

AMS. American Meteorological Society, 45 Beacon St., Boston, MA 02108-3693. (http://www.ametsoc.org/AMS/).

ASCE. American Society of Civil Engineers, 1801 Alexander Bell Dr., Reston, VA 20191-4400. (http://www.asce.org).

ASOS. Automated Surface Observing Station.

BAMS. Bulletin of the American Meteorological Society. See *AMS*.

Burn-in-place flare. A pyrotechnic device burned in a fixed location, such as the trailing edge of an aircraft wing.

CCN. Cloud condensation nuclei. Tiny solid and/or liquid (not predominantly pure water) *nuclei* upon which water vapor first condenses as the relative humidity approaches 100%. These *nuclei* are the seeds for cloud droplets.

Certified *Weather Modification* **Manager**. Certification of weather modification project managerial experience and skills granted by the WMA.

Certified *Weather Modification* **Operator**. Certification of weather modification project operational experience and skills granted by the WMA.

Cloud droplets. Liquid water droplets that are too small to precipitate since they typically range from a micrometer to tens of micrometers in diameter. Such droplets suspended in the atmosphere with other droplets form a cloud. Note: Human hair has an average thickness of about 65 micrometers.

CO_2. *See dry ice.*

Colloid. A mixture composed of two phases of matter, the dispersed phase and the continuous phase. It is an intermediate between suspensions and solutions. Colloids do not settle. In the case of fog, the dispersed phase is liquid and the continuous phase is air.

Continental *air mass*. A large body of air that forms over a continent. See *Air mass*.

Control area. A separate area, from the *target area* containing similar phenomena being modified (fog), wherein the natural processes can be monitored to provide a scientific standard used to assess the results from the application of unnatural stimuli (i.e., cloud seeding agent introduction) within fogs in the *target area*. Control areas may not be necessary for fog dispersal projects, especially since results of seeding are readily visible upon application. Nonetheless, control areas are still a good idea in fog dispersal projects for documenting the naturally changing physical characteristics of the fog being dispersed.

Dry ice. Frozen carbon dioxide (CO_2). Dry ice pellets have an equilibrium surface temperature of −78°C, and an operational temperature range colder than −2°C. Dry ice is used fairly often, especially in applications where environmental concerns about silver doses are heightened or the temperature is between 0°C and −5°C.

Evaporation-mixing fog. Fog that forms from the mixture of two different air parcels. Steam fog is an example of evaporation-mixing fog and may be seen over the thermal ponds at Yellowstone National Park, especially during the winter months (e.g., Dennis, 1980).

EWRI. Environmental Water Resources Institute, American Society of Civil Engineers, 1801 Alexander Bell Drive, Reston, VA 20191-4400. (http://www.EWRInstitute.org).

FAA. Federal Aviation Administration. The governmental entity that regulates aircraft operations, safety, and use of airways in the United States. Analogous entities exist in most other nations. (http://www.faa.gov).

Fog droplets. See *Cloud droplets.*

Glaciogenic. Causing the formation of ice.

Glaciogenic seeding. Seeding clouds with materials intended to increase and/or initiate the formation of ice crystals.

Ground fog. See *Radiation fog.*

Heterogeneous nucleation. The phase change of a substance to a more condensed state initiated by a nucleus with different physical and chemical properties than the substance. For example, the nucleation of ice crystals from supercooled water vapor using silver iodide as a nuclei. See *Nucleation.*

Homogeneous nucleation. The phase change of a substance to a more condensed state initiated by a *nuclei* with the same physical and chemical properties as the substance. For example, the nucleation of ice crystals from supercooled water vapor. *See nucleation.*

Hydrometeor. Cloud water or ice particles of any size and shape, either suspended in the air or precipitating.

Ice fog. A fog made up of small ice crystals, generally at air temperatures colder than −30°C.

Ice-forming nucleus. See ***IFN.***

Ice nucleus. Any particle that serves as a nucleus for the formation of ice crystals in the atmosphere. The subset of atmospheric particles (essentially all airborne matter) upon which ice crystals will form. These *nuclei* are typically water-insoluble particles and may be classified as hydrophobic condensation *nuclei*. These are sometimes abbreviated *IN* or IFN, which stand for ice-*nuclei* or ice-forming *nuclei*, respectively.

IFN. Ice-forming nucleus. See *Ice nucleus.*

IFR. Instrument flight rules. The FAA regulations pertaining to flight at altitudes of 18,000 feet (5.5 km) above mean sea level or higher over U.S. airspace or in any meteorological conditions necessitating the use of aircraft instrumentation for safe navigation.

IN. See *Ice nucleus.*

In situ measurements. In situ measurements refer to the gathering of information from within or on a media. That is, they are made in the actual location or environment of the object or entity measured. Such measurements are still the most common type of measurements, although *remote sensing measurements* are gaining on them.

JWM. *Journal of Weather Modification.* The official journal of the Weather Modification Association. See *WMA.*

KI. Chemical formula for potassium iodide.

Latent heat. The heat released or absorbed per unit mass by a system in a reversible, isobaric-isothermal change of phase. Simply, the heat released into the surrounding air or absorbed from the surrounding air when water changes its physical state. For example, the heat released when water vapor condenses is called latent heat of condensation; the heat released when a liquid water droplet freezes is called the latent heat of fusion.

Maritime *air mass*. A large body of air that forms over an ocean. See *Air mass.*

MSDS. Material safety data sheets.

NH_4I. Chemical formula for ammonium iodide.

NOAA. National Oceanic and Atmospheric Administration, U.S. Department of Commerce. The parent organization of the U.S. National Weather Service and the federal agency to which all U.S. Weather Modification activities must be reported. See www.noaa.gov.

Nucleation. Any process through which a phase change of a substance to a more condensed state is initiated upon a nucleus. The nucleus is either the vapor phase of the substance or a different substance.

Nuclei. A basis for future development and growth, a kernel. An airborne particle upon which water vapor molecules accumulate to form a droplet or ice crystal. See *CCN, Cloud condensation nuclei, Condensation nuclei, Ice-forming nuclei, Ice nuclei, IN, IFN.*

NWS. The National Weather Service. See *NOAA.*

Operational area. The operational area contains the target area and any surrounding upwind areas employed for treatment in order that the effect from seeding covers the target area, as well as for evaluating the project (excluding the control area as long as these areas have not had seeding conducted in them).

Opportunity recognition. The identification of those fogs capable of being dissipated using an appropriate fog dispersal method.

Project area. The area covered by all the equipment used for operations within all other areas used in evaluation or assessment by the contractors and/or sponsors.

Radiation. The transfer of heat energy without the involvement of a physical substance, such as air molecules. For example, the heat from a stove burner is an example of radiation because the heat would flow away from this burner regardless of whether we had air molecules, no air molecules (as in space), a body of water, water and a piece of metal, or a piece of metal between the burner and the receptor.

Radiation fog. A fog produced by radiative cooling of the earth's surface. Sometimes called a radiation or ground fog. *Radiation fog* that forms due to the atmospheric conditions in a valley may be termed valley fog. A fog that forms in a valley may not necessarily be called a valley fog. For example, fogs in the Central Valley of California are generally *radiation fog*s. See *Radiation.*

Radiosonde. An instrument package that senses and transmits temperature, dewpoint temperature, and pressure at various altitudes during its ascent into the atmosphere. These quantities can be converted into relative humidity. Radiosondes are carried aloft by weather balloons twice daily from many sites around the world and may also be employed by projects to bolster forecasting or research efforts. If their position is tracked in time, then they will also reveal the vertical wind field profile. The radiosonde that is tracked in time to also obtain wind field is known as a rawinsonde.

Remote sensing devices. Refers to one or more sensors on satellite, airborne, or ground-based platforms that either directly (active) or indirectly (passive) infer information about a property of an environment without being in or attached to that environment. Examples of ground-based remote sensing devices include, lidars, radars, radiometers, acoustic sounders, and sonic anemometers. Satellites have radiometers sensitive to one or more spectral regions of the electromagnetic spectrum onboard a satellite. Some of the latest satellite platforms include a radar, and will include a lidar in the near future. These kinds of devices can also be mounted on aircraft. Contrast with *In situ measurements.*

Remote sensing measurements. Refers to the measurements from *remote sensing devices.*

SAD. Seasonal affective disorder.

Seeded area. The area that is actually seeded. This may be the same as the *target area* for fog dispersal operations. See also *Target area.*

Seeding agents. Agents dispensed by any means in or near a cloud volume which are intended to modify the cloud characteristics. Remember, a cloud volume whose base is at the ground is a fog.

Silver iodide. A common *glaciogenic* seeding agent. Its chemical symbol is AgI. See *Ice nucleus.*

Steam fog. See *Evaporation-mixing fog.*

Stratus clouds. These clouds are uniformly stratified and nearly always have a uniform base. They are normally precipitation-free, but occasionally drizzle or light mist will fall out of stratus clouds. There are weather reports of light to moderate continuous rain from stratus clouds that form near an ocean.

Supercooled fog. A fog comprised of supercooled water droplets. See *Supercooled water.*

Supercooled water. Water that remains in the liquid state despite air temperatures colder than 0°C (32°F). Some nonpure water droplets may exist in a supercooled state to temperatures well below −40°C (−40°F) (e.g., Ohtake, 1992; Sassen, 1992; Young, 1993; Woodley and Rosenfeld, 1998; Black and Heymsfield, 2002).

Target area. The area where the effects from seeding operations are targeted. The target area is usually near the *control area* similar in character and climatology. The target area of a fog dispersal project is where the objective is to clear the fog or to increase the visibility to a distance that is suitable for the sponsor to conduct their restricted operations in that area. In the case of an airport, the target area commonly encompasses the runway plus 2 to 4 miles on each end. Equipment and instrumentation may be deployed within this area to gather data, monitor the operation, or seed *radiation fog*, since these are typically accompanied by relatively calm winds. Fogs accompanied by winds, such as advection fogs, could require data gathering, monitoring, or seeding that extends from the area covered for the *radiation fog* to upwind of the target area, depending on the meteorological conditions and other project scope considerations. See also *Control area, Seeded area.*

USAF. United States Air Force.

Visibility. The daytime distance at which a black target can just be detected against a horizon sky with a constant threshold of 2%. The daytime visibility depends inversely on the sum of the scattering cross sections of the target per unit volume (i.e., the number of fog droplets per unit volume). The scattering cross sections are proportional to the geometric cross sections of the fog droplets (i.e., their sizes).

Wind field. The vertical and horizontal values of wind speed and wind direction over a surface on the earth.

WMA. Weather Modification Association, P.O. Box 26926, Fresno, CA 93729-6926. (http://www.weathermodification.org).

WMO. World Meteorological Organization, 7 bis Avenue de la Paix, CP 2300-1211, Geneva 2-Switzerland. (http://www.wmo.ch/index-en.html).

7.0 REFERENCES

American Meteorological Society (AMS), 1998: Capability Statements. *Bull. Amer. Meteorol. Soc. (BAMS)*, **79**, 2771.

American Society of Civil Engineers (ASCE), 2003: ASCE Policy Statement #275. ATMOSPHERIC WATER RESOURCES MANAGEMENT. ASCE, Reston Va.

ASCE/EWRI Standard 42-04, 2004. *Standard Practice for the Design and Operation of Precipitation Augmentation Projects.* (Ed. D. Griffith). EWRI/ASCE Standard 42-04, ASCE, Reston Va., pp. 63.

ASRC, 1973: *ASPECTS'73.* Atmospheric Sciences Research Center (ASRC), SUNYA, Publication Number 324, pp. 29.

Benson, C.S., 1969: Ice Fog. Low Temperature Air Pollution. Geophys. Inst. Rept. UAG R-173, University of Alaska, (DDC No. AD631553), pp. 79.

Black, R.A., and G.M. Heymsfield, 2002: Partially frozen drops high in a hurricane. *BAMS-nowcast*, **83(June),** 844-844.

Blackmore III, W.H., 1992: A summary of the weather modification activities reported in the United States during 1990. *J. Weather Modification,* **24**, 130–131.

Blackmore III, W.H., 1994: A summary of the weather modification activities reported in the United States during 1992. *J. Weather Modification,* **26**, 129–130.

Bott, A., U. Silvers, and W. Zdunkowski, 1990: A radiation fog model with a detailed treatment of the interaction between radiative transfer and fog microphysics. *J. Atmos. Sci.,* **47**, 2153–2166.

Byers, H.R., 1959: *General Meteorology, 3rd Ed.* McGraw-Hill Book Company, Inc., New York, pp. 540.

Changnon, S., 1975: Present and future of weather modification. *BAMS*, **58,** 162–165.

Charak, M.T., 1981: Preliminary analysis of reported weather modification activities in the United States during 1980 and comparison with data for 1973–1979. *J. Weather Modification,* **13**, 231–234.

Cooper, C.P., 1977: *Ecological Implications of Weather Modification.* Paper prepared for the Weather Modification Advisory Board, U.S. Department of Commerce, pp. 19.

Cooper, C. P., and W. S. Jolly, 1969: *Ecological Effects of Weather Modification: Problem Analysis.* University of Michigan, Ann Arbor, School of Natural Resources, pp. 183.

DeFelice, T.P., 1998: *Introduction to Meteorological Instrumentation and Measurement.* Prentice Hall, New York/New Jersey. ISBN 0-13-243270-6, pp. 229.

DeMott, P.J., A.B. Super, G. Langer, D.C. Rogers, and J.T. McPartland, 1995: Comparative characterizations of the ice nucleus ability of AgI aerosols by three methods. *J. Weather Modification,* **27**, 1–16.

Dennis, A., 1980: *Weather Modification by Cloud Seeding.* Academic Press, New York, pp. 267.

Douglas, W.J., 1968: Toxic properties of material used in weather modification. *Proc. of 1st National Conference on Weather Modification,* State University of New York, Albany, N.Y.

Duynkerke, P.G., 1991: Radiation fog—a comparison of model simulation with detailed observations. *Mon. Wea. Rev.,* **119**, 324–341.

Ellrod, G.P., 1992: Potential applications of GOES-I 3.9 μm infrared imagery. *6th Conf. on Satellite Meteorology and Oceanography,* Jan 28–Feb 2, Atlanta, Ga. AMS, Boston, 184–187.

Ellrod, G.P., 1995: Advances in the detection and analysis of fog at night using GOES multispectral infrared imagery. *Wea. Forecasting,* **10**, 606–619.

EWRI/ASCE 39-03, 2003: *Standard Practice for the Design and Operation of Hail Suppression Projects.* EWRI/ASCE 39-03, ASCE, Reston Va., pp. 59.

Finnegan, W.G., and R.L. Pitter, 1997: Ion-induced charge separations in growing single ice crystals—Effects on growth and interaction processes. *J. Colloid and Interface Science,* **189**, 322–327.

Gazzi, M., V. Vicentini, and O. Vittori, 1975a: Basic theory of runway visual range evaluation in actual operations. *Rivista Italiana di Geofisica,* **11,** 3–8.

Gazzi, M., V. Vicentini, and O. Vittori, 1975b: Runway visual range evaluation during the 24 hour day. *Rivista Italiana di Geofisica,* **11**, 9–12.

George, J.J., 1951: Fog. In: *Compendium of Meteorology.* (Malone, T.F., Ed.). AMS, Boston, 1179–1189.

Griffith, D., M. Politovich, J.H. Renick, D.W. Reynolds, and D. Rogers, 1995: Cloud Seeding Modes and Instrumentation. Section 5, *Manual #81, Guidelines for Cloud Seeding to Augment Precipitation,* Ed. Kahan, A.M., D. Rottner, R. Sena, and C.G. Keyes, Jr.), ASCE, Reston, Va., 87–123.

Götz, G., E. Mészáros, and G. Vali, 1991: *Atmospheric Particles and Nuclei,* Akadémiai Kiadó, Budapest, pp. 274.

Heintzenberg, J., M. Wendisch, B. Yuskiewicz, D. Orsini, A. Wiedensohler, F. Stratmann, G. Frank, B.G. Martinsson, D. Schell, S. Fuzzi, and G. Orsi, 1998: Characteristics of haze, mist and fog. *Contr. Atmos. Phys (Beitr. Phys. Atmos.),* **71**, 21–31.

Hess, W. N., 1974: *Weather and Climate Modification.* John Wiley & Sons, New York, pp. 535.

Hindman, E. E., 1986: Characteristics of supercooled liquid water clouds at mountaintop sites in the Colorado Rockies. *J. Climate Appl. Meteorol.,* **17**, 49–63.

Holroyd, III, E.W., A.B. Super, and B.A. Silverman, 1978: The practicality of dry ice for on-top seeding of convective clouds. *J. Appl. Meteorol.,* **17**, 49–63.

Horn, R.D., W.G. Finnegan, and P.J. DeMott, 1983: Experimental studies of nucleation of dry ice. *J. Appl. Meteorol.,* **22**, 1567–1570.

Houze Jr., R. A., 1993: *Cloud Dynamics.* Academic Press, New York. pp. 595.

Huffman, P.J., and T. Ohtake, 1971: Formation and growth of ice fog particles at Fairbanks, Alaska. *J. Geophys. Res.,* **76(3)**, 657–665.

Kahan, A.M., D. Rottner, R. Sena, and C.G. Keyes, Jr., 1995: *Guidelines for Cloud Seeding to Augment Precipitation,* ASCE Manual of Professional Practice No. 81, American Society of Civil Engineers, Reston, Va., pp. 145.

Klein, D.A., 1978a: Environmental impacts of artificial ice nucleating agents. Dowden, Hutchington & Ross, Inc. Stroudsburg, Pa., pp. 218.

Klein, D.A., 1978b: Management of nucleating agents used in Weather Modification. Final report submitted to the National Science Foundation, ENV73-07821A05.

Kumai, 1982: Formation of ice crystals and dissipation of supercooled fog by artificial nucleation, and variations of crystal habit of early growth stages. *J. Appl. Meteorol.,* **21**, 579–587.

Malone, T.F., Ed., 1951: *Compendium of Meteorology.* AMS, Boston, pp. 1345.

Mason, B.J., 1981: The mechanisms of cloud seeding with dry ice. *J. Weather Modif.,* **13**, 11.

McFadden, T.T., 1976: Suppression of ice fog from cooling ponds. U.S. Army CRREL Report 76–43, Hanover, N.H., pp. 80.

McFadden, T.T., and C.M. Collins, 1978: Ice fog suppression using reinforced thin chemical films. U.S. Army CRREL Report 78-26, Hanover, N.H., pp. 25.

NRC (National Research Council), 2003. *Critical Issues in Weather Modification Research Report.* National Academies Press, Washington, D.C. (www.nap.edu/openbook/0309090539/html/R1.html), pp. 20.

Ohtake, T., 1992. Freezing points of H_2SO_4 aqueous solutions and formation of polar stratospheric clouds. In: *Nucleation and Atmospheric Aerosols, The Proceedings of the Thirteenth International Conference on Nucleation and Atmospheric Aerosols.* (Eds. Fukuta, N., Wagner, P.E.), 233–236.

Ohtake, T., and P. J. Huffman, 1969: Visual range in ice fog. *J. Appl. Meteorol.,* **8**, 499–501.

Pasquill, F., 1962: *Atmospheric Diffusion*, Van Nostrand, New York, pp. 209.

Pruppacher, H. R., and J. D. Klett, 1980: *Microphysics of Clouds and Precipitation.* D. Reidel Publishing Co., Boston, pp. 734.

Purdom, J.F.W., 1996: The use of NOAA's new generation of geostationary satellites to observe ocean phenomena. *8th Conference on Satellite Meteorology and Oceanography,* Jan 28–Feb 2, Atlanta, Ga, AMS, Boston, 96–99.

Rogers, R.R., and Yau, 1991: *A Short Course in Cloud Physics.* Pergamon Press, New York, pp. 307.

Sassen, K, 1992. Ice nuclei availability in the higher troposphere: Implications of a remote sensing cloud phase climatology. In: *Nucleation and Atmospheric Aerosols, The Proceedings of the Thirteenth International Conference on Nucleation and Atmospheric Aerosols.* (Eds. Fukuta, N., and Wagner, P.E.), 287–290.

Schaefer, V. J., 1946a: The production of clouds containing supercooled water droplets or ice crystals under laboratory conditions. *Bullet. Amer. Meteorol. Soc.,* **29**, 175–182.

Schaefer, V. J., 1946b: The production of ice crystals in a cloud of supercooled water droplets. *Science,* 104, 457–459.

Schaefer, V. J., 1948: The natural and artificial formation of snow in the atmosphere. EOS Trans., 20 (4), Am. Geophys. Union, 492–493.

Schaefer, V.J., 1951: *Snow and its relationship to experimental meteorology.* In: *Compendium of Meteorology.* (Malone, T.F., Ed.). AMS, Boston, 221–234.

Schaefer, V. J., 1953: Final Report for Project Cirrus. Part 1. Laboratory, Field and Flight Experiments. Contract DA-36-0390SC-15345, G. E. Report RL-785. General Electric Co., Schenectady, N.Y. 12306.

Schaefer, V. J., and J. A. Day, 1981: *A Field Guide to the Atmosphere.* Peterson Field Series, 26, Houghton Mifflin Co., Boston.

Schemenauer, R.S., 1998: Proceedings 1st International conference on Fog. Vancouver, B.C., Toronto University Press, Toronto, pp. 300.

Schemenauer, R.S., and P. Caselles, 1991: The quality of fog water collected for domestic and agricultural use in Chile. *J. Appl. Meteorol.,* **31**, 275–290.

Silverman, B.A., and A.I. Weinstein, 1974. Fog. In: *Weather and Climate Modification* (W.N. Hess, Ed.). John Wiley & Sons, New York.

Tomasi, C., and F. Tampieri, 1976a: Features of the proportionality coefficient in the relationship between visibility and liquid water content in haze and fog. *Atmosphere,* **14**, 61–76.

Tomasi, C., and F. Tampieri, 1976b: Size distribution models of small water droplets in mist and their volume extinction coefficients at visible and infrared wavelengths. *Atmospheric Environment,* **10**, 1005–1013.

Tonna, G., 1989: Physical and optical properties of fog at 74 wavelengths from 0.35 to 90 μm with a sample of 239 spectra. *J. Meteorol. Soc. Japan,* **67**, 5–8.

Vickers, W.W., and J.F. Church, 1966: Investigation of optimal design for supercooled cloud dispersal and techniques. *J. Appl. Meteor.,* **5**, 105–118.

Vonnegut, B., 1981: Misconception about cloud seeding with dry ice. *J. Weather Modification,* **13**, 9–11.

Welch, R.M., and B.A. Wielicki, 1986: The stratocumulus nature of fog. *J. Climate Appl. Meteorol.,* **25**, 101–111.

Welch, R.M., M.G. Ravichandran, and S.K. Cox, 1986: Prediction of quasi-periodic oscillations in *radiation fogs*. Part I—Comparisons of simple similarity approaches. *J. Atmos. Sci.,* **43**, 633–651.

Wendler, G., 1969: Heat balance studies during ice fog periods in Fairbanks, Alaska. *Monthly Weather Review,* **97**, 512.

Wetzel, M.A., R.D. Borys, and L.E. Xu, 1996: Satellite microphysical retrievals for land-based fog with validation by balloon profiling. *J. Appl. Meteorol.,* **35**, 810–829.

WMA (Weather Modification Association), 1986: Weather Modification Association weather modification capability statement. *J. Weather Modification,* **18**, WMA, Fresno, Ca., 141–142.

WMA (Weather Modification Association), 2003: Certification requirements. *J. Weather Modification,* **35**, WMA, Fresno, Ca., 74–75.

WMA (Weather Modification Association), 2004: Statement on standards and ethics for weather modification operators. *J. Weather Modification,* **36**, WMA, Fresno, Ca., 98–104.

WMO (World Meteorological Organization), 1992: WMO statement on the status of weather modification. *J. Weather Modification,* **25**, 1–6.

Woodley, W. L., and D Rosenfeld, 1998: Deep Convective Clouds with Sustained Supercooled Liquid Water Down to −37.5°C. *Nature,* **405**, 440–442.

Young, K.C., 1993: *Microphysical Processes in Clouds.* Oxford University Press, New York, pp. 427.

TABLE 8-1. Approved SI Prefixes

Factor	Name	Symbol
10^{-6}	micro	μ
10^{-3}	milli	m
10^{-2}	centi	c
10^{2}	hecto	h
10^{3}	kilo	k
10^{6}	mega	M

8.0 CONVERSION TABLES

Manuals of Standard Practice specify units in accordance with International System of Units, or SI. In the SI, the base unit of length is the meter (m); of mass, the kilogram (kg); of time, the second (s); and of temperature, the Kelvin (K). Derived SI units include those for area, the square meter (m^2); for volume, the cubic meter (m^3); and for speed, a meter per second ($m\ s^{-1}$). Other derived SI units relevant to this manual include force, the Newton (N, $kg\ m\ s^{-2}$); pressure, the Pascal (Pa, or $N\ m^{-2}$); and energy, the joule (J, or N m). The Celsius temperature scale (°C) is also a derived SI unit.

A number of prefixes are approved for use with SI units so that very large or very small quantities need not be displayed in unruly formats. That is, for example, 1,000 m can be expressed as 1 km. Likewise, one-millionth of a meter can be expressed as 1 μm. These prefixes are listed in Table 8-1.

There are a limited number of other units outside of the SI that are approved for use with the SI. These include minutes (min), hours (h), days (d), and liters (L). Some useful conversions are presented in Table 8-2, and some useful physical constant values are provided in Table 8-3.

TABLE 8-2. Some Useful Conversions

Quantity	SI Units	cgs Units	English Units
Density	$1\ kg\ m^{-3}$	$10^{-3}\ g\ cm^{-3}$	$6.24 \times 10^{-2}\ lb\ ft^{-3}$
Force[a]	$1\ kg\ m\ s^{-2}$	$10^{5}\ g\ cm\ s^{-2}$	0.224 lb–f (lb force)
Pressure[b]	$1\ kg\ m^{-1}\ s^{-2}$	$10\ g\ cm^{-1}\ s^{-2}$	0.021 lb-f ft^{-2}
Energy, work[c]	$1\ kg\ m^{2}\ s^{-2}$	$10^{7}\ g\ cm^{2}\ s^{-2}$	0.738 lb-f ft
Power[d]	$1\ kg\ m^{2}\ s^{-3}$	$10^{7}\ g\ cm^{2}\ s^{-3}$	0.738 lb-f ft s^{-1}
Temperature	1 °K	1 °C	1.8 °F
Heat or radiation flux [e]	1 W	$0.2389\ cal\ s^{-1}$	$3.414\ BTU\ h^{-1}$
Flux density or power per unit area	$1\ W\ m^{-2}$	$23.89 \times 10^{-6}\ cal\ cm^{-2}\ s^{-1}$	$0.317\ BTU\ ft^{-2}\ h^{-1}$
Latent heat	$1\ J\ kg^{-1}$	$0.2389 \times 10^{-3}\ cal\ g^{-1}$	$0.430 \times 10^{-3}\ BTU\ lb^{-1}$
Specific heat	$1\ J\ kg^{-1}\ °K^{-1}$	$0.2389 \times 10^{-3}\ cal\ g^{-1}\ °C^{-1}$	$0.2389 \times 10^{-3}\ BTU\ lb^{-1}\ °F^{-1}$
Thermal diffusivity	$1\ m^{2}\ s^{-1}$	$10^{4}\ cm^{2}\ s^{-1}$	$10.8\ ft^{2}\ s^{-1}$
Calorie	1.163×10^{-6} kW h		
Calorie	4.1868 J		

Adapted from DeFelice, 1998

[a] 1 Newton, N = $kg\ m\ s^{-2}$, 1 dyne = $1\ g\ cm\ s^{-2}$.

[b] 1 Pascal = $1\ kg\ m^{-1}\ s^{-2}$, $10\ g\ cm^{-1}\ s^{-2} = 10^{-2}$ mb.

[c] 1 Joule (J) = $1\ kg\ m^{2}\ s^{-2}$, 1 erg = $1\ g\ cm^{2}\ s^{-2}$. [d] 1 W = $1\ kg\ m^{2}\ s^{-3}$, 0.738 lb-f ft s^{-1} = 0.00134 hp (horsepower). [e] BTU = British thermal unit.

TABLE 8-3. Some Useful Constant Values

Physical Constants	
Universal gas constant	8.3145 J K^{-1} mol^{-1}
Boltzmann constant	1.381×10^{-23} J K^{-1} $molecule^{-1}$
Solar constant	1.37×10^{3} W m^{-2}
Dry air	
Gas constant for dry air	287.04 J K^{-1} kg^{-1}
Molecular weight of dry air	28.97 g mol^{-1}
Density of dry air (20°C, 1,000 mb)	1.209 kg m^{-3}
Specific heat at constant pressure	1.004×10^{3} J K^{-1} kg^{-1}
Thermal conductivity (0°C) (independent of pressure)	2.4×10^{-2} J m^{-1} K^{-1} s^{-1}
Dynamic viscosity (20°C)	1.818×10^{-5} kg m^{-1} s^{-1}
Mechanical equivalent of heat	4.184 J cal^{-1}
Standard atmospheric pressure	1.013×10^{2} kPa
Water substance	
Gas constant for water vapor	461 J K^{-1} kg^{-1}
Molecular weight of water	18.016 g mol^{-1}
Density of liquid water (0°C, 1,000 mb)	1.000×10^{3} kg m^{-3}
Density of ice (0°C, 1,000 mb)	0.917×10^{3} kg m^{-3}
Specific heat of liquid water (0°C)	4.218×10^{3} J K^{-1} kg^{-1}
Specific heat of water vapor at constant pressure	1.952×10^{3} J K^{-1} kg^{-1}
Specific heat of water vapor at constant volume	1.463×10^{3} J K^{-1} kg^{-1}
Latent heat of vaporization (0°C)	2.50×10^{6} J kg^{-1}
Latent heat of vaporization (100°C)	2.25×10^{6} J kg^{-1}
Latent heat of fusion (0°C)	3.34×10^{5} J kg^{-1}
Diffusivity of water vapor in air (20°C, 1,000 mb)	2.57×10^{-5} m^{2} s^{-3}

Adapted from DeFelice, 1998

INDEX